Solutions Manual
for Atkins and de Paula's Physical Chemistry for the Life Sciences

Maria Bohorquez
Drake University

With contributions from:
Krzysztof Kuczera, University of Kansas
Ronald Terry, Western Illinois University
James L. Pazun, Pfeiffer University

W. H. Freeman and Company
New York

Printed in the United States of America

ISBN-13: 978-0-7167-7262-0
ISBN-10: 0-7167-7262-0

Second Printing

W. H. Freeman and Company
41 Madison Avenue
New York, NY 10010
Houndmills, Basingstoke
RG21 6XS England

Table of Contents

Fundamentals

F.5 $Work = force \times distance = m \times g \times h$

$Work = (65\ \text{kg})\times(9.81\ \text{m s}^{-2})\times(3.5\ \text{m}) = 2.2\times10^{3}\ \text{N m}$

F.6 $E_{\text{K}} = 1/2\ mv^2$

$$E_K = \frac{1}{2}\times 58\ \text{g}\times\frac{1\ \text{kg}}{1000\ \text{g}}\times\left(30\ \text{m s}^{-1}\right)^2 = 26\ \text{J}$$

F.7 $E_{\text{K}} = ½mv^2$

$$v = \left(50\frac{\text{km}}{\text{h}}\right)\times\left(\frac{1\text{h}}{3600\ \text{s}}\right)\times\left(\frac{1000\ \text{m}}{1\ \text{km}}\right) = 14\ \frac{\text{m}}{\text{s}}$$

$$E_K = \frac{1}{2}\left(1.5\times10^{3}\text{kg}\right)\times\left(14\ \text{m s}^{-1}\right)^2 = 1.4\times10^{2}\,\text{kJ}$$

F.8 $E_{\text{K}} = 1/2\ mv^2$

$$E_{\text{K}} = \frac{1}{2}\times 29\ \text{g}\times\frac{1\ \text{kg}}{1000\ \text{g}}\times\left(400\ \text{m s}^{-1}\right)^2 = 2.32\ \ \text{kJ}$$

F.9 $E_{\text{P}} = mgh$

$$E_{\text{P}} = \left(\frac{0.20059\ \text{kg mol}^{-1}}{6.023\times10^{23}\,\text{atoms mol}^{-1}}\right)\times\left(9.81\ \text{m s}^{-2}\right)\times 0.760\ \text{m} = 2.48\times10^{-24}\,\text{J}$$

F.10 $E_P = mgh$

$$E_P = (25\ \text{g}) \times \left(\frac{1\ \text{kg}}{1000\ \text{g}}\right) \times (9.81\ \text{m s}^{-2}) \times (50\ \text{m}) = 12\ \text{J}$$

F.11

(a) 101.325 kPa = 760 torr,

$$\frac{110\ \text{kPa}}{101.325\ \text{kPa}} = \frac{x\ \text{torr}}{760\ \text{torr}}$$

825 torr

(b)1 atm = 1.01325 bar

$$\frac{0.997\ \text{bar}}{1.01325\ \text{bar}} = \frac{x\ \text{atm}}{1\text{atm}}$$

0.9839 atm

(c)1 atm = 101.325 kPa $= 1.01325 \times 10^5$Pa

$$\frac{2.15 \times 10^4\ \text{Pa}}{1.01325 \times 10^5\ \text{Pa}} = \frac{x\ \text{atm}}{1\ \text{atm}}$$

0.212 atm

(d)1 Torr = 133.32 Pa

$$\frac{723\ \text{Torr}}{1\ \text{Torr}} = \frac{x\ \text{Pa}}{133.32\ \text{Pa}}$$

9.64×10^4 Torr

F.12 $p = \rho g h$

$$p = (1.10 \times 10^3\ \text{kg m}^{-3}) \times (9.81\ \text{m s}^{-2}) \times (11.5 \times 10^3\ \text{m}) = 1.24 \times 10^8\ \text{Pa}$$

F.13

$$p = \rho g h$$

$$\frac{p(\text{Earth})}{p(\text{Mars})} = \frac{(9.81\ \text{m s}^{-2}) \times \rho \times h}{(3.7\ \text{m s}^{-2}) \times \rho \times h} = \frac{x}{0.0060\ \text{atm}}$$

$$p(\text{Earth}) = 0.016\ \text{atm}$$

F.14 $p = \rho g h$

(a) $p = (1.0\times10^3\text{kg m}^{-3})\times(9.81\ \text{m s}^{-2})\times(15\times10^{-2}\text{m}) = 1.5\times10^3\text{Pa}$

(b) $p = (1.0\times10^3\ \text{kg m}^{-3})\times(3.7\ \text{m s}^{-2})\times(15\times10^{-2}\text{m}) = 5.5\times10^2\ \text{Pa}$

F.15

$$p = 1\ \text{mmHg} = \rho g h = (1.35951\times10^4\text{kg m}^{-3})\times(9.80665\ \text{m s}^{-2})\times(10^{-3}\text{m})$$

$$= 133.322\ \text{kg m}^{-1}\text{s}^{-2} = 133.3223874\ \text{Pa}$$

$$(133.3223874\ \text{Pa})\times\left(\frac{1\ \text{torr}}{133.3223684\ \text{Pa}}\right) = 1.000000143\ \text{torr}$$

$$1\ \text{mmHg} = 1.000000143\ \text{Torr}$$

F.16 $\theta_{\text{Celsius}}/^\circ\text{C} = 5/9(\theta_{\text{Fahrenheit}}/^\circ\text{F} - 32)$

$\theta_{\text{Fahrenheit}}/^\circ\text{F} = 9/5(\theta_{\text{Celsius}}/^\circ\text{C}) + 32$

$\theta_{\text{Fahrenheit}}/^\circ\text{F} = 9/5(-273.15) + 32 = -459.67$

F.17 (a)

$$^\circ C = -209.9 + 0.141\times P = K - 273.15$$

$$P = \frac{(K - 273.15) + 209.9}{0.141}$$

(b)

$$^\circ C = -209.9 + 0.141\times P = \frac{F-32}{1.8}$$

$$P = \frac{F-32}{0.2538} + 1488$$

F.18 $pV = nRT$

$$n = \frac{pV}{RT} = \frac{(24.5\ \text{kPa})\times(0.250\ \text{L})}{(8.31447\ \text{L kPa K}^{-1}\ \text{mol}^{-1})\times(292.6\ \text{K})} = 2.52\times10^{-3}\ \text{mol}$$

F.19

$$pV = nRT$$

$$p = \frac{nRT}{V}$$

Mass of carbon dioxide =1.04 kg – 0.74 kg = 0.30 kg

$$n = \frac{3.0\times10^2\,g}{44.01\text{ g/mol}} = 6.8\text{ mol}$$

$$p = \frac{(6.8\text{ mol})\times(0.08205\ 74\text{ L atm K}^{-1}\text{ mol}^{-1})\times(293.15\text{ K})}{0.250\text{ L}} = 6.5\times10^2\text{atm}$$

F.20 $p_1V_1 = p_2V_2$

$$p_2 = \frac{V_1}{V_2}\times p_1$$

$$p_2 = \left(\frac{1.00\text{ L}}{0.100\text{ L}}\right)\times(1.00\text{ atm}) = 10\text{ atm}$$

F.21

$$\frac{P_1}{T_1} = \frac{P_2}{T_2}$$

$$P_2 = \frac{P_1}{T_1}\times T_2 = \left(\frac{125\text{ kPa}}{291.15\text{ K}}\right)\times 973.15 = 418\text{ kPa}$$

F.22 $p_1V_1 = p_2V_2$

$$p_2 = \frac{V_1}{V_2}\times p_1$$

$$p_2 = \left(\frac{7.20\text{ L}}{4.21\text{ L}}\right)\times(101\text{ kPa}) = 173\text{ kPa}$$

F.23

$$\frac{V_1}{T_1} = \frac{V_2}{T_2}$$

$$T_2 = \frac{V_2}{V_1} \times T_1 = \frac{V_1 + 0.14\ V_1}{V_1} \times T_1 = 1.14 \times T_1 = 387\ \text{K}$$

F.24 $\dfrac{p_1V_1}{T_1} = \dfrac{p_2V_2}{T_2}$

(a) $V_2 = \dfrac{p_1V_1T_2}{T_1p_2} = \dfrac{(104\ \text{kPa}) \times (2.0\ \text{m}^3) \times (268.1\ \text{K})}{(294.2\ \text{K}) \times (52\ \text{kPa})} = 3.6\ \text{m}^3$

(b) $V_2 = \dfrac{p_1V_1T_2}{T_1p_2} = \dfrac{(104\ \text{kPa}) \times (2.0\ \text{m}^3) \times (221.1\ \text{K})}{(294.2\ \text{K}) \times 880 \times (10^{-3}\ \text{kPa})} = 1.7 \times 10^2\ \text{m}^3$

F.25

$$p_1V_1 = p_2V_2$$

$$V_2 = \frac{p_1V_1}{p_1 + \rho gh}$$

$$V_2 = \frac{(1.01325 \times 10^5\,\text{Pa}) \times (3.0\ \text{m}^3)}{(1.01325 \times 10^5\,\text{Pa}) + (0.1025 \times 10^4\,\text{kg m}^{-3}) \times (9.81\ \text{m s}^{-2}) \times (50\ \text{m})} = 0.50\ \text{m}^3$$

F.26 $\dfrac{p_1V_1}{T_1} = \dfrac{p_2V_2}{T_2}$

$$V = \frac{4}{3}\pi r^3$$

$$V_1 = 4.1\ \text{m}^3;\ V_2 = 1.1 \times 10^2\ \text{m}^3$$

$$p_2 = \frac{p_1V_1T_2}{T_1V_2} = \frac{(1\ \text{atm}) \times (4.1\ \text{m}^3) \times (253\ \text{K})}{(293\ \text{K}) \times (1.1 \times 10^2\ \text{m}^3)} = 3.2 \times 10^{-2}\ \text{atm}$$

F.27 $M = \dfrac{\rho RT}{p} = \dfrac{(1.23\ \text{g L}^{-1})\times(8.314\ 47\ \text{L kPa K}^{-1}\text{mol}^{-1})(330\ \text{K})}{25.5\ \text{kPa}} = 132\ \text{g mol}^{-1}$

F.28 $c = \left(\dfrac{3RT}{M}\right)^{1/2}$

(a)

(i) $c = \left(\dfrac{3\times(8.31447\ \text{J K}^{-1}\ \text{mol}^{-1})\times(77\ \text{K})}{(4\times10^{-3}\ \text{kg mol}^{-1})}\right)^{1/2} = 693\ \text{m s}^{-1}$

(ii) $c = \left(\dfrac{3\times(8.31447\ \text{J K}^{-1}\text{mol}^{-1})\times(298\ \text{K})}{(4\times10^{-3}\text{kg mol}^{-1})}\right)^{1/2} = 1363\ \text{m s}^{-1}$

(iii) $c = \left(\dfrac{3\times(8.31447\ \text{J K}^{-1}\ \text{mol}^{-1})\times(1000\ \text{K})}{(4\times10^{-3}\ \text{kg mol}^{-1})}\right)^{1/2} = 2497\ \text{m s}^{-1}$

(b)

(i) $c = \left(\dfrac{3\times(8.31447\ \text{J K}^{-1}\ \text{mol}^{-1})\times(77\ \text{K})}{(16\times10^{-3}\ \text{kg mol}^{-1})}\right)^{1/2} = 346\ \text{m s}^{-1}$

(ii) $c = \left(\dfrac{3\times(8.31447\ \text{J K}^{-1}\ \text{mol}^{-1})\times(298\ \text{K})}{(16\times10^{-3}\ \text{kg mol}^{-1})}\right)^{1/2} = 681\ \text{m s}^{-1}$

(iii) $c = \left(\dfrac{3\times(8.31447\ \text{J K}^{-1}\ \text{mol}^{-1})\times(1000\ \text{K})}{(16\times10^{-3}\ \text{kg mol}^{-1})}\right)^{1/2} = 1248\ \text{m s}^{-1}$

F.29

$$e = \int_0^\infty s f(s)\mathrm{d}c$$

$$e = 4\pi\left(\frac{m}{2\pi k_B T}\right)^{3/2} \int_0^\infty s^3 \mathrm{e}^{-ms^2/2k_\mathrm{B}T}\mathrm{d}s$$

$$e = 4\pi\left(\frac{m}{2\pi k_\mathrm{B} T}\right)^{3/2} \times \frac{1}{2a^2}$$

$$a = \frac{m}{2k_B T}$$

$$e = 4\pi\left(\frac{m}{2\pi k_B T}\right)^{3/2} \times \frac{1}{2}\left(\frac{2k_B T}{m}\right)^2$$

$$e = \sqrt{\frac{8k_B T}{\pi m}} = \sqrt{\frac{8RT}{\pi M}}$$

F.30

$$c_\mathrm{rms} = \left(c^{-2}\right)^{1/2}$$

$$c^{-2} = \int_0^\infty c^2 f(c)\mathrm{d}c$$

$$c^{-2} = 4\pi\left(\frac{m}{2\pi k_B T}\right)^{3/2} \int_0^\infty c^4 \mathrm{e}^{-mc^2/2k_B T}\mathrm{d}c$$

$$c^{-2} 4\pi\left(\frac{m}{2\pi k_B T}\right)^{3/2} \times \frac{3}{8a^2}\sqrt{\frac{\pi}{a}}$$

$$a = \frac{m}{2k_B T}$$

$$c^{-2} = 4\pi\left(\frac{m}{2\pi k_B T}\right)^{3/2} \times \frac{3}{8\left(\frac{m}{2k_B T}\right)^2} \times \sqrt{\frac{\pi}{\frac{m}{2k_B T}}}$$

$$\overline{c^2} = 4\pi\left(\frac{m}{2\pi k_B T}\right)^{3/2} \times \frac{3\sqrt{\pi}}{8} \times \left(\frac{2k_B T}{m}\right)^{5/2}$$

$$\overline{c^2} = \frac{3k_B T}{m}\left(\frac{N_A}{N_A}\right) = \frac{3RT}{M}$$

$$c_{\mathrm{rms}} = \sqrt{\frac{3RT}{M}}$$

F.31

$$f(c) = 4\pi\left(\frac{m}{2\pi k_B T}\right)^{3/2} s^2 e^{-ms^2/2k_B T}$$

$$\frac{\mathrm{d}f(c)}{\mathrm{d}c} = 4\pi\left(\frac{m}{2\pi k_B T}\right)^{3/2}\left[2se^{-ms^2/2k_B T} + s^2 e^{-ms^2/2k_B T}\left(-\frac{2ms}{2k_B T}\right)\right]$$

$$\frac{\mathrm{d}f(c)}{\mathrm{d}c} = 4\pi\left(\frac{m}{2\pi k_B T}\right)^{3/2} se^{-ms^2/2k_B T}\left(2 - \frac{ms^2}{k_B T}\right) = 0$$

$$4\pi\left(\frac{m}{2\pi k_B T}\right)^{3/2} se^{-ms^2/2k_B T}\left(2 - \frac{ms^2}{k_B T}\right) = 0$$

$$\left(2 - \frac{ms^2}{k_B T}\right) = 0$$

$$s^2 = \frac{2k_B T}{m}$$

$$s = \sqrt{\frac{2RT}{M}}$$

F.32 $f = F(s)\,\Delta s$, where: $F(s) = 4\pi\left(\dfrac{M}{2\pi RT}\right)^{3/2} s^2 \exp\left(\dfrac{-Ms^2}{2RT}\right)$, and,

Δs is the range of speeds = 10 m/s.
Using s = 300 m/s, and T = 500 K:

$$f = F(s)\Delta s$$

$$= 4\pi\left(\frac{.028\text{ kg/mol}}{2\pi(8.314\text{ J K}^{-1}\text{ mol}^{-1})(500\text{ K})}\right)^{3/2} \times$$

$$(300\text{ m/s})^2 \exp\left(\frac{-(.028\text{ kg mol}^{-1})(300\text{ m s}^{-1})^2}{2\times(8.314\text{ J K}^{-1}\text{mol}^{-1})(500\text{ K})}\right)\times(10\text{ m/s})$$

$$f = 0.0093$$

Chapter 1: The First Law

1.8 $Work = mgh$

$$Work = (0.200\ \text{kg}) \times (9.81\ \text{m s}^{-2}) \times (20\ \text{m}) = 39\ \text{J}$$

1.9 $C_6H_{12}O_6(s) + 6\ O_2(g) \rightarrow 6\ CO_2(g) + 6\ H_2O(l)$

(a) $w = -p_{ex}\Delta V = -p_{ex}(V_f - V_i) = -p_{ex}(V_{CO_2} - V_{O_2})$

Work of expansion is zero because final and initial gaseous volumes are the same.

(b) The combustion of 1.0g (5.5×10^{-3} mol) of glucose requires 0.033 mol oxygen and produces 0.033 mol carbon dioxide and 0.033 mol water vapor. In this case, the formation of water vapor contributes to the final volume.

Final and initial volumes are calculated by:

$$V_f = \frac{0.066RT}{p_{ex}}$$

$$V_i = \frac{0.033RT}{p_{ex}}$$

Work of expansion is then:

$$w = -0.033\ RT = -(0.033\ \text{mol}) \times (8.314\,47\ \text{J K}^{-1}\ \text{mol}^{-1}) \times (293.15\ \text{K}) = -80\ \text{J}$$

1.10 $Work = -p_{ex}\,\Delta V = -p_{ex}\,Ah$

$$Work = -(100\times10^3\text{Pa})\times(100\ \text{cm}^2) \times \left(\frac{1\ \text{m}}{100\ \text{cm}}\right)^2 \times (10.0\ \text{cm}) \times \left(\frac{1\ \text{m}}{100\ \text{cm}}\right)$$

$$Work = -100\ \text{Pa m}^3 = -100\ \text{J}$$

1.11 (a)

$$w = -p_{\text{ex}}(V_{\text{f}} - V_{\text{i}}) = -\left(30.0\times10^{3}\,\text{Pa}\right)\times\left(3.3\times10^{-3}\,\text{m}^{3}\right) = -99\ \text{J}$$

(b)

$$w = -nRT\ln\frac{V_{\text{f}}}{V_{\text{i}}}$$

$$= (-0.281\ \text{mol})\times(8.314\,47\ \text{J K}^{-1}\text{mol}^{-1})\times(310\ \text{K})\times\ln\frac{16}{12.7} = -167\ \text{J}$$

1.12 (a)

$$\text{d}w = -\frac{nR}{V}(T_{\text{i}} - cV + cV_{\text{i}})\text{d}V = -\frac{nRT_{\text{i}}}{V}\text{d}V + nRc\text{d}V - \frac{\text{n}RcV_{\text{i}}}{V}\text{d}V$$

$$\text{d}w = -nR\left(T_{\text{i}} + cV_{\text{i}}\right)\frac{\text{d}V}{V} + nRc\text{d}V$$

$$w = -nR\left(T_{\text{i}} + cV_{\text{i}}\right)\ln\frac{V_{\text{f}}}{V_{\text{i}}} + nRc(V_{\text{f}} - V_{\text{i}})$$

(b) As shown in Exercise 1.13, the graph of work done by the system against the final volume suggests work is greater than isothermal expansion if c takes a negative value.

1.13

$$\text{d}w = -\frac{nR}{V}(T_{\text{i}} - cV + cV_{\text{i}})\text{d}V = -\frac{nRT_{\text{i}}}{V}\text{d}V + nRc\text{d}V - \frac{nRcV_{\text{i}}}{V}\text{d}V$$

$$\text{d}w = -nR\left(\text{T}_{\text{i}} + cV_{\text{i}}\right)\frac{\text{d}V}{V} + nRc\text{d}V$$

$$w = -nR\left(\text{T}_{\text{i}} + cV_{\text{i}}\right)\ln\frac{V_{\text{f}}}{V_{\text{i}}} + nRc(V_{\text{f}} - V_{\text{i}})$$

Work versus the final volume:

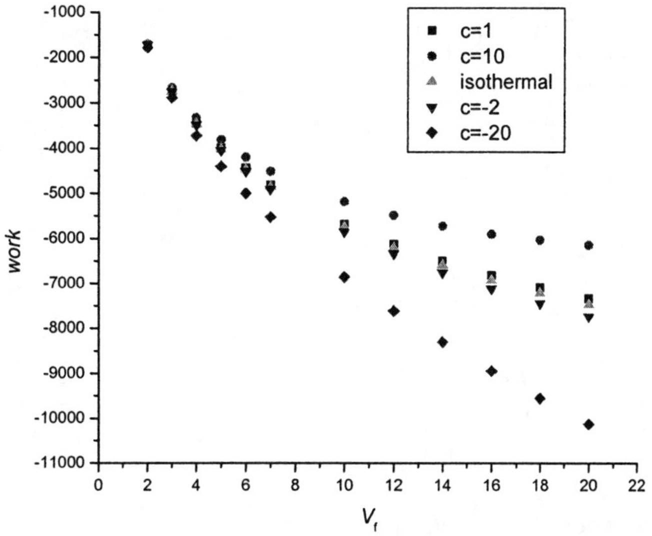

1.14 $q = nC_{P,\mathrm{m}}\Delta T$

We need to calculate moles of air:

$$n = \frac{pV}{RT} = \frac{(101\ \mathrm{kPa})\times(55\ \mathrm{dm})\times(65\ \mathrm{dm})\times(30\ \mathrm{dm})}{(8.31447\ \mathrm{L\ kPa\ K^{-1}\ mol^{-1}})\times(283\ \mathrm{K})} = 4.6\times10^{3}\ \mathrm{mol}$$

$$q = (4.6\times10^{3}\ \mathrm{mol})\times(21\ \mathrm{J\ K^{-1}})\times(10\ \mathrm{K}) = 9.7\times10^{2}\ \mathrm{kJ}$$

$$t = (9.7\times10^{2}\ \mathrm{kJ})\ /\ (1.5\ \mathrm{kW}) = 6.5\times10^{2}\ \mathrm{s}$$

1.15

$$w = -nRT\ln\frac{V_\mathrm{f}}{V_\mathrm{i}} = -(1\ \mathrm{mol})\times(8.31447\ \mathrm{J\ K^{-1}mol^{-1}})\times(300\ \mathrm{K})\times\ln\frac{30}{22} = -774\ \mathrm{J}$$

$$q = -w = +774\ \mathrm{J}$$

1.16 *Work* = *mgh*

Work = $(0.200\ \mathrm{kg})\times(9.81\ \mathrm{m\ s^{-2}})\times(1.55\ \mathrm{m}) = 3.04\ \mathrm{J}$

The work done by the animal is −3.04 J. The change in internal energy is then:

$$\Delta U = w + q$$

$$\Delta U = (-3.0\ \mathrm{J}) + (-5.0\ \mathrm{J}) = -8.0\ \mathrm{J}$$

1.17 It does not change.

1.18 $\Delta H = q_P = 1.2 \text{ kJ}$

$q = nC_{P,\text{m}}\Delta T$

$$\frac{q}{\Delta T} \approx C_P \approx \frac{(1.2\times10^3 \text{ J})}{15 \text{ K}} = 80 \text{ J K}^{-1}$$

1.19 (a)

$$C_{P,\text{m}} - C_{V,\text{m}} =$$

$$= \left(\frac{\partial H}{\partial T}\right)_P - \left(\frac{\partial U}{\partial T}\right)_V = \left(\frac{\partial U}{\partial T}\right)_P + \left(\frac{\partial (PV)}{\partial T}\right)_P - \left(\frac{\partial U}{\partial T}\right)_V$$

$$= \left(\frac{\partial U}{\partial T}\right)_P + P\left(\frac{\partial V}{\partial T}\right)_P - \left(\frac{\partial U}{\partial T}\right)_V$$

The relationship between $\left(\frac{\partial U}{\partial T}\right)_P$ and $\left(\frac{\partial U}{\partial T}\right)_V$ is:

$$dU = \left(\frac{\partial U}{\partial T}\right)_V dT + \left(\frac{\partial U}{\partial V}\right)_T dV$$

$$\left(\frac{\partial U}{\partial T}\right)_P = \left(\frac{\partial U}{\partial T}\right)_V + \left(\frac{\partial U}{\partial V}\right)_T \left(\frac{\partial V}{\partial T}\right)_P$$

Substitute $\left(\frac{\partial U}{\partial T}\right)_P$ into $C_{P,\text{m}} - C_{V,\text{m}}$

$$C_{P,\text{m}} - C_{V,\text{m}} = \left[\left(\frac{\partial U}{\partial V}\right)_T + P\right]\left(\frac{\partial V}{\partial T}\right)_P = 0 + nR$$

(b)

$q = nC_{P,\text{m}}\Delta T$

$$C_{P,\text{m}} = \frac{229 \text{ J}}{(3 \text{ mol})\times(2.06 \text{ K})} = 37 \text{ J K}^{-1} \text{ mol}^{-1}$$

$$C_{V,\text{m}} = (37 \text{ J K}^{-1}\text{mol}^{-1}) - (8.31447 \text{JK}^{-1}\text{mol}^{-1}) = 29 \text{ J K}^{-1} \text{ mol}^{-1}$$

1.20 (a) $q = nC_{P,\text{m}}\Delta T$

$q_P = \Delta H_\text{m}$

$\Delta H_\text{m} = (37 \text{ J K}^{-1} \text{ mol}^{-1}) \times (22 \text{ K}) = 8.1\times10^2 \text{ J mol}^{-1}$

(b) $\Delta U_\text{m} = \Delta H_\text{m} - \Delta(pV_\text{m}) = \Delta H_\text{m} - R\Delta T$

$\Delta U_\text{m} = (8.1\times10^2 \text{ J mol}^{-1}) - (8.31447 \text{ J K}^{-1} \text{ mol}^{-1}) \times (22 \text{ K}) = 6.3\times10^2 \text{ J mol}^{-1}$

1.21

$$U_m = a + bT + cT^2$$

$$\left(\frac{\partial U_m}{\partial T}\right)_V = C_{V,m} = b + 2Tc$$

1.22 The integration of $dH = C_{P,m}dT$ is below:

$$dH = \int_{T_1}^{T_2}\left(a + bT + \frac{c}{T^2}\right)dT$$

$$dH = a\int_{T_1}^{T_2} dT + b\int_{T_1}^{T_2} TdT + c\int_{T_1}^{T_2}\frac{1}{T^2}dT$$

$$\Delta H = a(T_2 - T_1) + \frac{b}{2}(T_2^2 - T_1^2) - c\left(\frac{1}{T_2} - \frac{1}{T_1}\right)$$

The substitution of *a*, *b*, *c*, initial and final temperature yields

$$\Delta H = (44.22\ \mathrm{J\ K^{-1}\ mol^{-1}}) \times (310\ \mathrm{K} - 288\ \mathrm{K}) +$$

$$\frac{(8.79\times10^{-3}\ \mathrm{J\ K^{-2}\ mol^{-1}})}{2}\left((310\ \mathrm{K})^2 - (288\ \mathrm{K})^2\right) +$$

$$+ (8.62\times10^{5}\ \mathrm{J\ K\ mol^{-1}}) \times \left(\frac{1}{310\mathrm{K}} - \frac{1}{288\mathrm{K}}\right)$$

$$\Delta H = 818\ \mathrm{J\ mol^{-1}}$$

1.23 (a) Linear

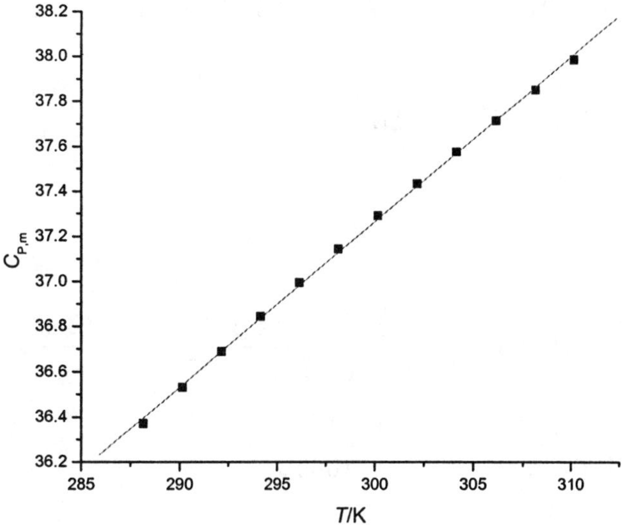

$$\Delta H = a\left(T_{\mathrm{f}} - T_{\mathrm{i}}\right) + \frac{b}{2}\left(T_{\mathrm{f}}^2 - T_{\mathrm{i}}^2\right) - c\left(\frac{1}{T_{\mathrm{f}}} - \frac{1}{T_{\mathrm{i}}}\right)$$

(b)

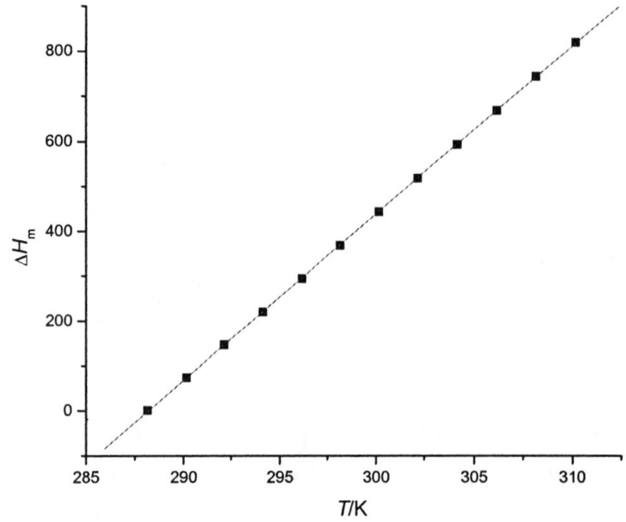

1.24 (a) exothermic, (b) endothermic, (c) endothermic, (d) endothermic, (e) endothermic

1.25 (a)

C(s, graphite) + $O_2 \rightarrow CO_2$ $\qquad \Delta H^{\circ} = -393.5 \text{kJmol}^{-1}$

$CO_2 \rightarrow$ C(s, diamond) + O_2 $\qquad \Delta H^{\circ} = +395.41 \text{ kJmol}^{-1}$

Overall:

C(s, graphite) $\rightarrow$ C(s, diamond) $\qquad \Delta H^{\circ} = +1.91 \text{ kJmol}^{-1}$

(b)

$$\Delta H = \Delta U + P\Delta V$$

$$\Delta U = \Delta H - P\Delta V$$

$$\Delta U = 1.91 \text{ kJ} - \left[\begin{array}{l}\left(150\times10^3 \text{bar}\times\dfrac{10^5 \text{Pa}}{1 \text{ bar}}\right)\times\left(\dfrac{12 \text{ g}}{3.510 \; g \text{ mL}^{-1}} - \dfrac{12 \text{ g}}{2.250 \text{ g mL}^{-1}}\right) \\ \times\left(\dfrac{\text{L}}{1000 \text{ mL}}\times\dfrac{10^{-3}\text{m}^3}{\text{L}}\right)\times\dfrac{1 \text{ kJ}}{1000 \text{ J}}\end{array}\right]$$

$$\Delta U = 30.62 \text{ kJ}$$

1.26 $q = nC_{P,\mathrm{m}}\Delta T$

$$\Delta T = \frac{q}{nC_{p,\mathrm{m}}} = \frac{(10\times10^6\ \mathrm{J})}{(65\times10^3\ \mathrm{g})\times\left(\frac{1\ \mathrm{mol}}{18\ \mathrm{g}}\right)\times(75.29\ \mathrm{J\ K^{-1}mol^{-1}})} = 37\ \mathrm{K}$$

Since the enthalpy of vaporization of water is 40.7 kJ mol^{-1},

$\frac{(10\times10^6\ \mathrm{J})}{(40.7\times10^3\ \mathrm{J\ mol^{-1}})} = 2.5\times10^2$ mol water molecules will evaporate, given that 10 MJ of energy are produced by a typical human body. The mass of water that should be evaporated is then 4.5 kg.

1.27

$$q = \Delta_{\mathrm{fus}}H^{-} + C_{\mathrm{P,l}}\Delta T + \Delta_{\mathrm{vap}}H^{-}$$

$$q = \left((6.01\times10^3\ \mathrm{J\ mol^{-1}})\times\frac{100\ \mathrm{g}}{18\ \mathrm{g\ mol^{-1}}}\right) +$$

$$\big((100\ \mathrm{g})\times(4.18\ \mathrm{J\ K^{-1}g^{-1}})\times(100\ \mathrm{K})\big) + \left((40.7\times10^3\ \mathrm{J\ mol^{-1}})\times\frac{100\ \mathrm{g}}{18\ \mathrm{g\ mol^{-1}}}\right)$$

$$q = 301\ \mathrm{kJ}$$

1.28 The number of broken and formed bonds for the combustion of glucose and decanoic acid are shown below:

$C_6H_{12}O_6(s) + 6\ O_2(g) \rightarrow 6\ CO_2(g) + 6\ H_2O(l)$

Number and type of broken bonds	5 C-C	5 O-H	7 C-O	7 C-H	6 O=O	
Number and type of formed bonds		12 O-H				12 C=O

$C_{10}H_{20}O_2(s) + 14\ O_2(g) \rightarrow 10\ CO_2(g) + 10\ H_2O(l)$

Number and type of broken bonds	9 C-C	1 O-H	1 C-O	19 C-H	1 C=O	
Number and type of formed bonds		20 O-H				20 C=O

The number of stronger bonds such as C=O and O-H is greater for decanoic acid, which justifies the higher specific enthalpy compared with glucose. (By ‘lower’ in this case, we mean “more negative,” or larger in magnitude but negative).

1.29 (a) $C_6H_{12}O_6(aq) \rightarrow 2\ CH_3CH(OH)COOH(aq)$

Number and type of broken bonds	5 C-C	5O-H	7C-O	7C-H	
Number and type of formed bonds	4C-C	4O-H	4C-O	8C-H	2C=O
Energy(kJ/mol)	348	463	3×360	-412	-2×743

$\Delta H = -7$ kJ mol^{-1} (Compare to the experimental value of -121 kJ mol^{-1})

(b) $C_6H_{12}O_6(s) + 6\ O_2(g) \rightarrow 6\ CO_2(g) + 6\ H_2O(l)$

Number and type of broken bonds	5 C-C	5 O-H	7 C-O	7 C-H	6 O=O	
Number and type of formed bonds		12 O-H				12 C=O
Energy(kJ/mol)	5×348	-7×463	7×360	7×412	6×497	-12×743

$\Delta H = -2{,}031$ kJ mol^{-1} (Compare to the experimental value of -2808 kJ mol^{-1})

1.30 (a) Combustion of sucrose in air releases -5645 kJ mol^{-1}

$C_{12}H_{22}O_{11}(s) + 12\ O_2(g) \rightarrow 12\ CO_2(g) + 11\ H_2O(l)$

$$\frac{1.5\ \text{g}}{342\ \text{g mol}^{-1}} = (4.4\times10^{-3}\ \text{mol})$$

$$q = (-5645\ \text{kJ mol}^{-1})\times(4.4\times10^{-3}\text{mol}) = -25\ \text{kJ}$$

(b) Assuming a human body with a mass of 65 kg

$Work = mgh$

$$h = \frac{Work}{mg} = \frac{0.25\times\left(25\times10^{3}\ \text{J}\right)}{(65\ \text{kg})\times\left(9.81\ \text{m s}^{-2}\right)} = 9.8\ \text{m}$$

(c) Combustion of glucose in air releases -2808 kJ mol^{-1}

$C_6H_{12}O_6(s) + 6\ O_2(g) \rightarrow 6\ CO_2(g) + 6\ H_2O(l)$

$$\frac{2.5\ \text{g}}{180\ \text{g mol}^{-1}} = 1.4\times10^{-2}\ \text{mol}$$

$$q = (-2808\ \text{kJ mol}^{-1})\times(1.4\times10^{-2}\ \text{mol}) = -39\ \text{kJ}$$

(d) Assuming a human body with a mass of 65 kg

$Work = mgh$

$$h = \frac{Work}{mg} = \frac{0.25\times(39\times10^3\ \text{J})}{(65\ \text{kg})\times(9.81\ \text{m s}^{-2})} = 15\ \text{m}$$

1.31 (a) The sum of the following reactions:

$C_3H_8(g) + 5O_2(g) \rightarrow 3\ CO_2(g) + 4\ H_2O(l)$ $\quad \Delta H^{-} = -2220$ kJ

$C_3H_8(l) \rightarrow C_3H_8(g)$ $\quad \Delta H^{-} = +15$ kJ

allows the calculation of the standard enthalpy of combustion of liquid propane:

$C_3H_8(l) + 5O_2(g) \rightarrow 3\ CO_2(g) + 4\ H_2O(l)$ $\quad \Delta H^{-} = -2205$ kJ

(b)

$$\Delta H = \Delta U + P\Delta V$$

$$\Delta U = \Delta H - \Delta(nRT)$$

$$\Delta U = \Delta H + 2RT$$

$$\Delta U = -2205\ \text{kJ} + [2\times(8.31447\ \text{J K}^{-1}\text{mol}^{-1})\times(298.15\ \text{K})]\times(10^{-3}\frac{\text{kJ}}{\text{J}})$$

$$\Delta U = -2200\ \text{kJ}$$

1.32 (a) The standard enthalpy of the combustion of ethane is -1560 kJ mol^{-1}.

(b) The specific enthalpy of the combustion of ethane is 52 kJ g^{-1}.

(c) The specific enthalpy of CH_4 is -55 kJ/g. By comparing specific enthalpies of combustion, we find that ethane is a less efficient fuel than methane, since it provides less energy per gram.

1.33 $-286\ \text{kJ mol}^{-1} + 285.830\ \text{kJ mol}^{-1} = -0.17\ \text{kJ mol}^{-1}$

1.34 (a)

$$\Delta_r H^\circ = 2\times\Delta_f H^\circ\ (\text{glycine, aq}) - \Delta_f H^\circ\ (\text{H}_2\text{O, l}) - \Delta_f H^\circ\ (\text{glycine–glycine, aq})$$

$$\Delta_r H^\circ = (-2\times469.8 - 285.83 + 747.7)\ \text{kJ mol}^{-1}$$

$$\Delta_r H^\circ = -477.7\ \text{kJ mol}^{-1}$$

(b)

$$\Delta_r H^\circ = 6\times\Delta_f H^\circ\ (\text{CO}_2,\ \text{g}) + 6\times\Delta_f H^\circ\ (\text{H}_2\text{O, l}) - \Delta_f H^\circ\ (\text{fructose, aq})$$

$$\Delta_r H^\circ = (-2361 - 1715 + 1265.6)\ \text{kJ mol}^{-1}$$

$$\Delta_r H^\circ = -2810\ \text{kJ mol}$$

(c)
$\Delta_r H^{\ominus} = \Delta_f H^{\ominus}(NO,g) + \Delta_f H^{\ominus}(O,g) - \Delta_f H^{\ominus}(NO_2,g)$
$\Delta_r H^{\ominus} = (+90.25 + 249.17 - 33.18)\ \text{kJ mol}^{-1} = 306.24\ \text{kJ mol}^{-1}$
Note: (c) only occurs in the atmosphere via absorption of light.

1.35 The standard enthalpy of combustion of pyruvic acid can be calculated using the following equations:
$2\ CH_3COCOOH(s) + 2\ H_2O(l) \rightarrow C_6H_{12}O_6(s) + O_2(g)$ $\Delta_r H^{\ominus} = 480.7\ \text{kJ mol}^{-1}$
$C_6H_{12}O_6(s) + 6\ O_2(g) \rightarrow 6\ CO_2(g) + 6\ H_2O(l)$ $\Delta_r H^{\ominus} = -2808\ \text{kJ mol}^{-1}$
$CH_3COCOOH(s) + 5/2\ O_2(g) \rightarrow 3\ CO_2(g) + 2\ H_2O(l)$ $\Delta_r H^{\ominus} = -1163$
The standard enthalpy of the formation of pyruvic acid is calculated by adding the following combustion reactions:

$3\ CO_2(g) + 2\ H_2O(l) \rightarrow CH_3COCOOH(s) + 5/2\ O_2(g)$
$\Delta_r H^{\ominus} = +1163\ \text{kJ mol}^{-1}$
$3\ C(s, \text{graphite}) + 3\ O_2 \rightarrow 3\ CO_2\ (g)$
$\Delta_r H^{\ominus} = -1182\ \text{kJ mol}^{-1}$
$2\ H_2(g) + O_2(g) \rightarrow 2\ H_2O(l)$
$\Delta_r H^{\ominus} = -572\ \text{kJ mol}^{-1}$
$3\ C(s, \text{graphite}) + 3/2\ O_2(g) + 2\ H_2(g) \rightarrow CH_3COCOOH(s)$
$\Delta_r H^{\ominus} = -591\ \text{kJ mol}^{-1}$

1.36 (a) We estimate the enthalpy of denaturation of the protein at two different temperatures by assuming the change in heat capacities is independent of temperature over the temperature range of interest.
$\Delta_r H^{\ominus}(T_2) = \Delta_r H^{\ominus}(T_1) + \Delta_r C_p^{\ominus} \Delta T$
(i) $\Delta_r H^{\ominus}(351\ \text{K}) = (217.6\ \text{kJ mol}^{-1}) + (6.3\ \text{kJ K}^{-1}\ \text{mol}^{-1}) \times (53\ \text{K}) = 5.5 \times 10^2\ \text{kJ mol}^{-1}$
(ii) $\Delta_r H^{\ominus}(263\ \text{K}) = (217.6\ \text{kJ mol}^{-1}) - (6.3\ \text{kJ K}^{-1}\ \text{mol}^{-1}) \times (35\ \text{K}) = -2.9\ \text{kJ mol}^{-1}$
(b) No

1.37 $\Delta_{vap} H = C_{P,m}(H_2O,l)(-75\ \text{K}) + 44.01\ \text{kJ} + C_{P,m}(H_2O,g)(75\ \text{K}) = +40.88\ \text{kJ mol}^{-1}$

1.38 Higher: $\Delta H \sim \Delta C_{p,m} \Delta T$
If $\Delta C_{p,m}$ is positive, ΔH will be less negative (smaller in magnitude), if $\Delta C_{p,m}$ is negative, ΔH will be more negative (larger in magnitude).

1.40 A more accurate form of Kirchhoff's law could be derived if the following integration is carried out:

$$\int_{T_1}^{T_2} \mathrm{d}H = \int_{T_1}^{T_2} \Delta_r C_P^{\ominus} \mathrm{dT} = \int_{T_1}^{T_2} \left(a + bT + \frac{c}{T^2} \right) \mathrm{d}T$$

$$\int_{T_1}^{T_2} \mathrm{d}H = a\int_{T_1}^{T_2} \mathrm{d}T + b\int_{T_1}^{T_2} T\mathrm{d}T + c\int_{T_1}^{T_2} \frac{1}{T^2}\mathrm{d}T$$

$$\Delta_r H(T_2) = \Delta_r H(T_1) + a(T_f - T_i) + \frac{b}{2}(T_f^2 - T_i^2) - c\left(\frac{1}{T_f} - \frac{1}{T_i} \right)$$

1.39

$$\Delta_r U(T') = \Delta_r H(T') - \Delta(nRT')$$

$$\Delta_r U(T') = (\Delta_r H(T) + \Delta_r C_P \times \Delta T) - \Delta(nRT')$$

Chapter 2: The Second Law

2.6 $\Delta S_{\text{sur}} = \dfrac{q_{\text{sur}}}{T}$

$$\Delta S_{\text{sur}} = \frac{120\ \text{J}}{293\ \text{K}} = +0.410\ \text{J K}^{-1}$$

2.7 The combustion of 100 g of glucose gives off 1560 kJ.

$$\Delta S = \frac{q}{T} = 5.03 \times 10^{3}\ \text{J K}^{-1}$$

2.8 (a) $\Delta S = \dfrac{q}{T}$

The amount of heat for a reversible isothermal expansion or contraction is $nRT \ln \dfrac{V_2}{V_1}$, so the entropy change due to reversible isothermal expansion is

$$\Delta S = nR \ln \frac{V_2}{V_1}.$$

(b) $\Delta S_{\text{m}} = (8.31447\ \text{J K}^{-1}\text{mol}^{-1}) \times \ln \dfrac{4.5\ \text{L}}{1.5\ \text{L}} = +9.1\ \text{J K}^{-1}\text{mol}^{-1}$

(c) The amount of carbon dioxide (in moles) is:

$$n = \frac{pV}{RT} = \frac{(1.00\ \text{atm}) \times (15.0\ \text{L})}{(8.20574 \times 10^{-2}\ \text{L atm K}^{-1}\text{mol}^{-1}) \times (250\ \text{K})} = 0.731\ \text{mol}$$

$$\Delta S_{\text{m}} = (0.731\ \text{mol}) \times (8.31447\ \text{J K}^{-1}\text{mol}^{-1}) \times \ln \frac{\text{V}_2}{15.0\ \text{L}} = -10.0\ \text{J K}^{-1}$$

$$V_2 = \text{e}^{-\left(\frac{(10.0\ \text{J K}^{-1})}{(0.731\ \text{mol}) \times (8.31447\ \text{J K}^{-1}\text{mol}^{-1})}\right)} \times 15.0\ \text{L} = 2.89\ \text{L}$$

2.9

$$(a)\Delta S = \frac{q}{T} = +0.12\ \text{kJ K}^{-1}$$

$$(b)\Delta S = \frac{q}{T} = -0.12\ \text{kJ K}^{-1}$$

2.10 The change in entropy of melting 100 g of ice at 0°C is:

$$\Delta_{\text{fus}}S = \frac{\Delta_{\text{fus}}H(T_{\text{fus}})}{T_{\text{fus}}} = \frac{(6010\ \text{J mol}^{-1})\times\left(\dfrac{100\ \text{g}}{18\ \text{g mol}^{-1}}\right)}{273.15\ \text{K}} = +122\ \text{J K}^{-1}$$

The entropy change associated with heating 100 g of water to 100°C is:

$$\Delta S = C_P \ln\frac{T_2}{T_1} = (75.29\ \text{J K}^{-1}\text{mol}^{-1})\times\left(\frac{100\ \text{g}}{18\ \text{g mol}^{-1}}\right)\times\ln\frac{373.15\ \text{K}}{273.15\ \text{K}}$$

$$\Delta S = +130\ \text{J K}^{-1}$$

The entropy change of vaporization is:

$$\Delta_{\text{vap}}S = \frac{\Delta_{\text{vap}}H(T_{\text{b}})}{T_{\text{b}}} = \frac{(4070\ \text{J mol}^{-1})\times\left(\dfrac{100\ \text{g}}{18\ \text{g mol}^{-1}}\right)}{373.15\ \text{K}} = +606\ \text{J K}^{-1}$$

The total entropy change of melting and evaporating 100 g of ice is $+858\ \text{J K}^{-1}$.

(a) T constant through step 1, then increases linearly, then constant.
(b) H increases linearly throughout.
(c) S abruptly increases for step 1, increases logarithmically (curved, concave down), then abruptly increases.

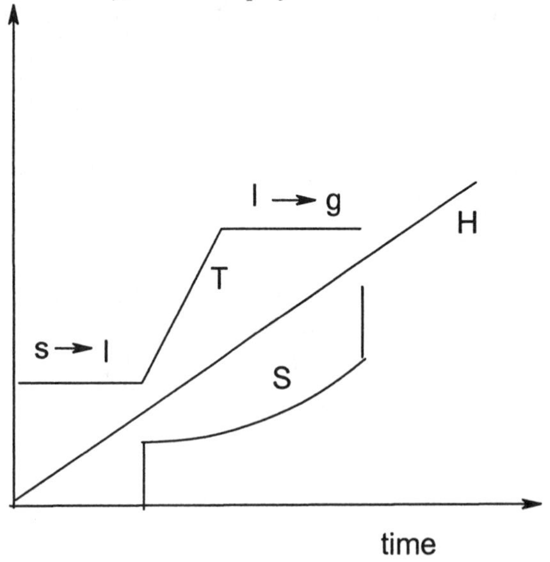

2.11 $\Delta S = n \times C_{P,\text{m}} \ln \frac{T_\text{f}}{T_\text{i}} = 23.6 \text{ J K}^{-1}$

2.12 We can use Equation 2.9b to estimate the molar entropy of potassium chloride.

$$S_\text{m}(T) = 1/3 C_{V,\text{m}}(T) = \frac{1}{3} \times (1.2 \text{ mJ K}^{-1} \text{ mol}^{-1}) = +4 \times 10^{-4} \text{ J K}^{-1} \text{ mol}^{-1}$$

2.13

$$dq = T dS = C dT$$

$$dS = C \frac{dT}{T} = (a + bT + \frac{a}{T^2}) \frac{dT}{T} = \frac{a}{T} dT + b dT + \frac{a}{T^3} dT$$

$$\Delta S = a \ln \frac{T_\text{f}}{T_\text{i}} + b(T_\text{f} - T_\text{i}) - \frac{a}{2}\left(\frac{1}{T_\text{f}^2} - \frac{1}{T_\text{i}^2}\right)$$

2.14 We need to calculate the final temperature of the system. Since the system is an insulated vessel, the heat given off by water at 80° C is the same as the heat absorbed by water at 10° C.

$$q_{80^\circ\text{C}} = -q_{10^\circ\text{C}}$$

$$\left(\frac{100 \text{ g}}{18 \text{ g mol}^{-1}}\right) \times (75.5 \text{ J K}^{-1} \text{ mol}^{-1}) \times (T_\text{f} - 80^\circ\text{C})$$

$$= -\left(\frac{100 \text{ g}}{18 \text{ g mol}^{-1}}\right) \times (75.5 \text{ J K}^{-1} \text{ mol}^{-1}) \times (T_\text{f} - 10^\circ\text{C})$$

$$(T_\text{f} - 80^\circ\text{C}) = -(T_\text{f} - 10^\circ\text{C})$$

$$T_\text{f} = 45^\circ\text{C}$$

The entropy change of water at 80°C is:

$$\Delta S = C_P \ln \frac{T_2}{T_1} = (75.5 \text{ J K}^{-1}\text{mol}^{-1}) \times \left(\frac{100 \text{ g}}{18 \text{ g mol}^{-1}}\right) \times \ln \frac{318.15 \text{ K}}{353.15 \text{ K}}$$

$$\Delta S = -43.7 \text{ J K}^{-1}$$

The entropy change of water at 10°C is:

$$\Delta S = C_P \ln \frac{T_2}{T_1} = (75.5 \text{ J K}^{-1} \text{ mol}^{-1}) \times \left(\frac{100 \text{ g}}{18 \text{ g mol}^{-1}}\right) \times \ln \frac{318.15 \text{ K}}{283.15 \text{ K}}$$

$$\Delta S = +48.9 \text{ J K}^{-1}$$

The total entropy change is $+5.20 \text{ J K}^{-1}$.

2.15 The entropy of unfolding of lysozyme at 25.0°C can be calculated using the following paths:

(i) Heating the folded protein from 25.0°C to the transition temperature.

(ii) Unfolding at the transition temperature.

(iii) Cooling of the unfolded protein to 25.0°C.

$$\Delta S = \left(C_{P,\text{f}} \ln \frac{348.65}{298.15}\right) + \left(\frac{509 \text{ kJ mol}^{-1}}{348.65 \text{ K}}\right) - \left(C_{P,\text{u}} \ln \frac{348.65}{298.15}\right)$$

$$\Delta S = \ln \frac{348.65}{298.15} \times \left(C_{P,\text{f}} - C_{P,\text{u}}\right) + \frac{509 \text{ kJ mol}^{-1}}{348.65 \text{ K}}$$

$$\Delta S = \ln \frac{348.65}{298.15} \times \left(6.28 \text{ kJ mol}^{-1} \text{ K}^{-1}\right) + \frac{509 \text{ kJ mol}^{-1}}{348.65 \text{ K}}$$

$$\Delta S = 2.44 \times 10^{3} \text{ J K}^{-1}$$

2.16 The entropy of transition from graphite to diamond at 2000 K is:

$$\Delta_{\text{trs}} S = \frac{\Delta_{\text{trs}} H(T_{\text{trs}})}{T_{\text{trs}}} = \frac{190 \text{ J mol}^{-1}}{2000 \text{ K}} = +0.95 \text{ J K}^{-1} \text{ mol}^{-1}$$

2.17 (a)

$$\Delta S = \frac{\Delta_{\text{vap}} H}{T_{\text{b}}} = \frac{35.27 \text{kJ mol}^{-1}}{337.25 \text{ K}} = 0.1046 \text{ kJ K}^{-1} \text{mol}^{-1}$$

(b)

$-0.1046 \text{ kJ K}^{-1} \text{mol}^{-1}$

2.18 (a) The experimental observation that the entropy of vaporization for several different liquids is the same, indicates a similar increase in disorder. This finding suggests that most liquids have a similar degree of disorder near the boiling point in the absence of strong intermolecular interactions. The same can be said for gases near the boiling point, which is definitely the case in the limit that they are behaving ideally.

(b) According to Trouton's rule, the entropy of vaporization of liquids at the boiling point is about $85 \text{ J K}^{-1} \text{ mol}^{-1}$. Using this value, we can calculate the enthalpy of the vaporization of octane:

$$\Delta_{\text{vap}} H = \Delta_{\text{vap}} S \times T_{\text{b}} = (85 \text{ J mol}^{-1}) \times (399 \text{ K}) = +34 \text{ kJ K}^{-1} \text{ mol}^{-1}$$

(c) Water has a greater vaporization than predicted by Trouton's rule. The extensive network of hydrogen bonds renders water rather ordered relative to most other liquids – this is neglected in Trouton's rule. Thus, the degree of disorder generated in water during vaporization is much larger than in other liquids.

2.19 The entropy of fusion at 25.0°C can be calculated using the following paths:

(i) Heating of the compound in the solid phase from 25.0°C to 146.0°C

(ii) Phase transition of the compound from solid to liquid at 146.0°C

(iii) Cooling of the compound in the liquid phase to 25.0°C

$$\Delta S = \left(C_{P,s} \ln\frac{419}{298}\right) + \left(\frac{32 \text{ kJ mol}^{-1}}{419 \text{ K}}\right) - \left(C_{P,l} \ln\frac{419}{298}\right)$$

$$\Delta S = \ln\frac{419}{298} \times \left(19 \text{ J mol}^{-1} \text{ K}^{-1} - 28 \text{ J mol}^{-1} \text{ K}^{-1}\right) + \frac{32 \times 10^{3} \text{ J mol}^{-1}}{419 \text{ K}}$$

$$\Delta S = \ln\frac{419}{298} \times \left(-9 \text{ J mol}^{-1} \text{ K}^{-1}\right) + \frac{32 \times 10^{3} \text{ J mol}^{-1}}{419 \text{ K}}$$

$$\Delta S = 0.34 \times \left(-9 \text{ J mol}^{-1} \text{ K}^{-1}\right) + 76 \text{ J mol}^{-1} \text{ K}^{-1}$$

$$\Delta S = 0.34 \times \left(-9 \text{ J mol}^{-1} \text{ K}^{-1}\right) + 76 \text{ J mol}^{-1} \text{ K}^{-1}$$

$$\Delta S = 73 \text{ J mol}^{-1} \text{ K}^{-1}$$

2.20 The standard reaction entropy at 298 K can be calculated using equation 2.10

$$\Delta_r S^{\ominus} = 2 \times S_m^{\ominus}(C_2H_5OH, l) + 2 \times S_m^{\ominus}(CO_2, g) - S_m^{\ominus}(C_6H_{12}O_6, s)$$

$$\Delta_r S^{\ominus} = \left(321.4 \text{ J K}^{-1}\text{mol}^{-1}\right) + \left(427.4 \text{ J K}^{-1} \text{ mol}^{-1}\right) - \left(212 \text{ J K}^{-1} \text{ mol}^{-1}\right)$$

$$\Delta_r S^{\ominus} = 537 \text{ J K}^{-1} \text{ mol}^{-1}$$

2.21 (a)

$$\Delta G = \Delta H - T\Delta S$$

$$\Delta G = -125 \text{ kJ mol}^{-1} + \left(310.15\ K \times 0.126 \text{ kJ K}^{-1}\text{mol}^{-1}\right)$$

$$\Delta G = -85.9 \text{ kJ mol}^{-1}$$

The reaction is spontaneous.

(b)

$$\Delta G = -T\Delta S_{\text{total}}$$

$$\Delta S_{total} = \frac{85.9 \text{ kJ mol}^{-1}}{310.15 \text{ K}} = +0.277 \text{ kJ K}^{-1}\text{mol}^{-1}$$

2.22 At constant temperature and pressure $\Delta G = w'_{\text{max}}$

To climb through 10 m the amount of energy required is

$$w'_{\text{max}} = (65 \text{ kg}) \times (9.81 \text{ m s}^{-2}) \times (10 \text{ m}) = 6.4 \text{ kJ}$$

2828 kJ of energy are available per mole of glucose, so:

$$(6.4 \text{ kJ})/(2828 \text{ kJ/mol}) = 2.3 \times 10^{-2} \text{ mol}$$

2.3×10^{-3} mol are needed to climb through 10 m. (0.41 g)

2.23 (a)

glutamate + NH_4^+=glutamine $\Delta G = 14.2$ kJ mol^{-1}

ATP=ADP +P_i $\Delta G = -31$ kJ mol^{-1}

Glutamate + NH_4^+ +ATP= glutamine +ADP + P_i $\Delta G = -16.8$ kJ mol^{-1}

The hydrolysis will drive the formation of glutamine.

(b) 14.2 / 31 = 0.46 mol ATP.=

2.24 The Gibbs energy of hydrolysis of ATP is -31 kJ mol^{-1}. The minimum number of ATP molecules that would be needed is 42/31 = 1.35 mol = 8.15×10^{23}.

2.25 The number of ATP mol hydrolyzed each second is 1.66×10^{-18} mol. A typical cell will deliver $(1.66\times 10^{-18}$ mol$)\times$ $(31\times 10^3$ J mol$^{-1}) = 5.15\times 10^{-14}$ W per second. If the volume of a cell is 4.18×10^{-15} m^3, the power density of the cell is 12.3 W m^{-3}. The battery has the greater power density (150,000 W m^{-3}).

Chapter 3: Phase Equilibria

3.7 (a)

$$\Delta G_{\mathrm{m}} = V_{\mathrm{m}}\Delta p$$

$$\Delta G_{\mathrm{m}} = (17.5\times10^{-3}\,\mathrm{L})\times\frac{10^{-3}\,\mathrm{m}^3}{1\mathrm{L}}\times\left((0.103\times10^4\,\mathrm{kg\ m^{-3}})\times(9.81\ \mathrm{m\ s^{-2}})\times(11.5\times10^3\,\mathrm{m})\right)$$

$$\Delta G_{\mathrm{m}} = +2.03\ \mathrm{kJ\ mol^{-1}}$$

(b)

$$\Delta G_{\mathrm{m}} = V_{\mathrm{m}}\Delta p$$

$$\Delta G_{\mathrm{m}} = (14.7\times10^{-3}\,\mathrm{L})\times\frac{10^{-3}\,\mathrm{m}^3}{1\mathrm{L}}\times\begin{bmatrix}\left((1.36\times10^4\ \mathrm{kg\ m^{-3}})\times(9.81\ \mathrm{m\ s^{-2}}\times0.760\ m)\right)- \\ (160\times10^{-3})\end{bmatrix}$$

$$\Delta G_{\mathrm{m}} = +1.49\ \mathrm{J\ mol^{-1}}$$

3.8

$$\Delta G_{\mathrm{m}} = V_{\mathrm{m}}\Delta p$$

$$V_{\mathrm{m}} = \left(\frac{891.5\ \mathrm{g\ mol^{-1}}}{0.95\ \mathrm{g\ mL^{-1}}}\right)\times\left(\frac{1\ \mathrm{L}}{1000\ \mathrm{mL}}\right) = 0.94\ \mathrm{L\ mol^{-1}}$$

$$\Delta G_{\mathrm{m}} = (0.94\times10^{-3}\ \mathrm{m^3\ mol^{-1}})\times(9.81\ \mathrm{m\ s^{-2}})\times(1.03\times10^3\ \mathrm{kg\ m^{-3}})\times(2000\ \mathrm{m}) = 19\ \mathrm{kJ\ mol^{-1}}$$

3.9 (a)

$$\Delta G_{\mathrm{m}} = RT\ln\frac{p_{\mathrm{f}}}{p_{\mathrm{i}}}$$

$$\Delta G_{\mathrm{m}} = (8.31447\ \mathrm{J\ K^{-1}mol^{-1}})\times(293.15\ \mathrm{K})\times\ln 2$$

$$\Delta G_{\mathrm{m}} = +1.7\ \mathrm{kJ\ mol^{-1}}$$

(b)

$$\Delta G_{\mathrm{m}} = RT \ln \frac{p_{\mathrm{f}}}{p_{\mathrm{i}}}$$

$$\Delta G_{\mathrm{m}} = (8.31447 \text{ J K}^{-1}\text{mol}^{-1}) \times (293.15 \text{ K}) \times \ln(0.00027 \times 1.01325)$$

$$\Delta G_{\mathrm{m}} = -20 \text{ kJ mol}^{-1}$$

3. 10 As shown in Equation 3.3, the change in molar Gibbs energy at constant pressure is $\Delta G_{\mathrm{m}} = -S_{\mathrm{m}}\Delta T$.

Because molar entropy is always positive, the slopes of a graph G_{m} versus T are negative for the solid, liquid and gas phases. Since $S_{\mathrm{m}}(\mathrm{g}) > S_{\mathrm{m}}(\mathrm{l}) > S_{\mathrm{m}}(\mathrm{s})$, the slopes for the different phases follow that trend.

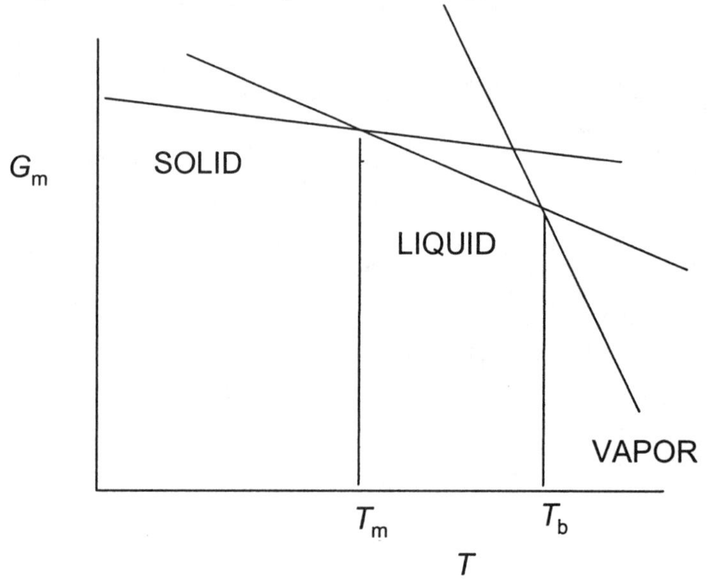

3.11 The volume of the room is 1.0176×10^5 L. The mass of each substance that will be found in the room can be calculated assuming ideal behavior of the vapor:

(a)

$$n = \frac{(3.2 \text{ kPa}) \times (1.0176 \times 10^5 \text{ L})}{(8.31447 \text{ L kPa K}^{-1}\text{mol}^{-1}) \times (298.15 \text{ K})}$$

mass = 2.37 kg

(b)

$$n = \frac{(14 \text{ kPa}) \times (1.0176 \times 10^5 \text{ L})}{(8.31447 \text{ L kPa K}^{-1}\text{mol}^{-1}) \times (298.15 \text{ K})}$$

mass = 44.8 kg

(c)

$$n = \frac{(0.23\times10^{-3}\ \text{kPa})\times(1.0176\times10^{5}\,\text{L})}{(8.31447\ \text{L kPa K}^{-1}\text{mol}^{-1})\times(298.15\ \text{K})}$$

$$\text{mass} = 1.89\ \text{kg}$$

3.12 Yes, because the temperature and pressure conditions are below the triplet point of water. The partial pressure of water that will ensure the frost remains would have to be greater than 611 Pa (4 Torr).

3.13 (a) Water in the vapor phase will change first to liquid, and then to solid.
(b) The time is proportional to the amount of heat lost from water, so the slope $\mathrm{d}T/\mathrm{d}t$ of the cooling curve is proportional to the reciprocal of the water's heat capacity. The slope of the cooling curve for the solid is twice that of the liquid. The slope of the cooling curve for the vapor is 1.125 times greater that of the solid.

3.14 The water will freeze, and exist in the solid phase.

3.15 (a) The value of the enthalpy term is assumed to be proportional to the number of hydrogen bonds formed. Upon unfolding, non-polar residues will become exposed to water molecules, which will re-organize in structured cages around non-polar residues. The hydrogen bonding network of water changes as evidenced by the entropy values.
(b) At T_m the change in Gibbs energy is zero, which leads to:

$$(n-4)\Delta H_\text{m} = (n-2)T_\text{m}\Delta S_\text{m}$$

$$T_\text{m} = \left(\frac{n-4}{n-2}\right)\frac{\Delta H_\text{m}}{\Delta S_\text{m}}$$

(c) A plot of $T_m/(\Delta_{hb}H_m/\Delta_{hb}S_m)$ against n is shown below:

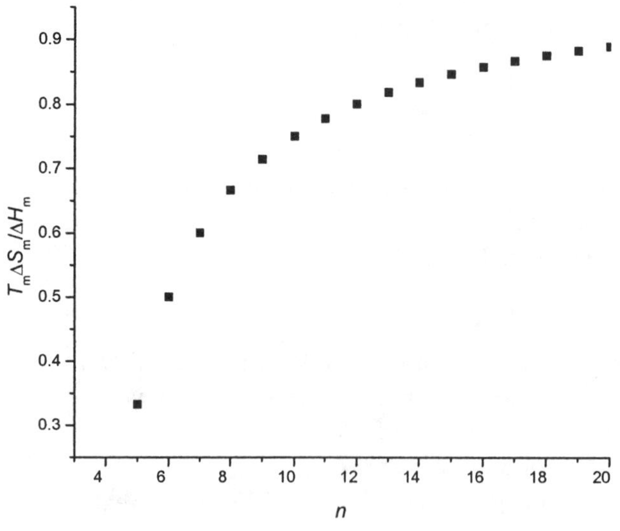

$n = 17$

3.16 (a) $\Delta_{init}G^{-}$ is positive, because by bringing together the two strands to form DNA, there should be a loss of entropy. $\Delta_{seq}G^{-}$ is negative because the values are related to the number of hydrogen bonds and π stacking interactions between base pairs.

(b) The standard Gibbs energy change can be estimated using the $\Delta_{seq}G^{-}$ provided.

$$\Delta_{DNA}G^{-} = \Delta_{init}G^{-} + \sum \Delta_{seq}G^{-}(\text{sequences})$$

$$\Delta_{DNA}G^{-} = +7.5 \text{ kJ mol}^{-1} + \begin{pmatrix} 2\times\Delta_{seq}G^{-}(\text{AG/TC}) + \Delta_{seq}G^{-}(\text{GC/CG}) \\ + \Delta_{seq}G^{-}(\text{TG/AC}) \end{pmatrix} \text{kJ mol}^{-1}$$

$$\Delta_{DNA}G^{-} = +7.5 \text{ kJ mol}^{-1} + (-10.8 - 10.5 - 6.7) \text{kJ mol}^{-1}$$

$$\Delta_{DNA}G^{-} = -20.5 \text{ kJ mol}^{-1}$$

In a similar manner, the standard enthalpy and entropy changes are estimated below:

$$\Delta_{\text{DNA}}H^{\ominus} = \sum \Delta_{\text{seq}}H^{\ominus}(\text{sequences})$$

$$\Delta_{\text{DNA}}H^{\ominus} = \Delta_{\text{init}}H^{\ominus} + \begin{pmatrix} 2 \times \Delta_{\text{seq}}H^{\ominus}(\text{AG/TC}) + \Delta_{\text{seq}}H^{\ominus}(\text{GC/CG}) + \\ \Delta_{\text{seq}}H^{\ominus}(\text{TG/AC}) \end{pmatrix} \text{kJ mol}^{-1}$$

$$\Delta_{\text{DNA}}H^{\ominus} = -127 \text{ kJ mol}^{-1}$$

$$\Delta_{\text{DNA}}S^{\ominus} = \sum \Delta_{\text{seq}}S^{\ominus}(\text{sequences})$$

$$\Delta_{\text{DNA}}S^{\ominus} = \Delta_{\text{init}}S^{\ominus} + \begin{pmatrix} 2 \times \Delta_{\text{seq}}S^{\ominus}(\text{AG/TC}) + \Delta_{\text{seq}}S^{\ominus}(\text{GC/CG}) + \\ \Delta_{\text{seq}}S^{\ominus}(\text{TG/AC}) \end{pmatrix} \text{J K}^{-1} \text{ mol}^{-1}$$

$$\Delta_{\text{DNA}}S^{\ominus} = -357 \text{ J K}^{-1} \text{ mol}^{-1}$$

(c) Figure 3.17 shows that the melting temperature changes linearly with the fraction of G-C base pairs: $T_{\text{m}}/\text{K} = 325 + 39.7f$.
An estimate on the melting temperature of the piece of DNA shown in part (b) is:
$T_{\text{m}}/\text{K} = 325 + 39.7 \times 0.6$
$T_m = 348 \text{ K}$
The fraction of G-C base pairs in this example is 3/5.

3.17 (760 Torr) – (47 Torr) = 713 Torr

3.18 According to Dalton's law, the pressure exerted by a mixture of perfect gases is the sum of the pressures that each gas would exert if it were alone in the container at the same temperature.
(a) The volume is calculated using the information about nitrogen.
There is 225 mg of nitrogen, so:

$$n_{\text{N}_2} = \frac{0.225 \text{ g}}{28.013 \text{ g mol}^{-1}} = 8.04\times10^{-3} \text{ mol}$$

$$V = \frac{nRT}{p} = \frac{(8.04\times10^{-3}\text{mol})\times(8.31447 \text{ L kPa K}^{-1}\text{mol}^{-1})\times(300 \text{ K})}{15.2 \text{ kPa}}$$

$V = 1.32 \text{ L}$
(b) The total pressure of the mixture can be calculated using Equation 3.6, $p_{\text{N}_2} = x_{\text{N}_2}p$.
The number of moles for each gas in the mixture besides nitrogen can be found:

$$n_{CH_4} = \frac{0.320\text{ g}}{16.04\text{ g mol}^{-1}} = 1.99\times10^{-2}\text{ mol}$$

$$n_{Ar} = \frac{0.175\text{ g}}{39.95\text{ g mol}^{-1}} = 4.38\times10^{-3}\text{ mol}$$

The mole fraction of nitrogen in the mixture is:

$$x_{N_2} = \frac{n_{N_2}}{n_{N_2}+n_{Ar}+n_{CH_4}} = \frac{8.04\times10^{-3}\text{ mol}}{\left((8.04\times10^{-3}) + (4.38\times10^{-3}) + (1.99\times10^{-2})\right)\text{ mol}}$$

$$x_{N_2} = 0.248$$

The total pressure of the mixture is then:

$$p = \frac{p_{N_2}}{x_{N_2}} = \frac{15.2\text{ kPa}}{0.248} = 61.3\text{ kPa}$$

3.19 (a)

$$n = (0.2500\text{ L})\times(0.112\text{ mol L}^{-1})$$

$$\text{mass} = 5.04g$$

(b)

$$n = 0.250\text{kg}\times0.112\text{ mol kg}^{-1}$$

$$\text{mass} = 5.04g$$

3.20 Following Example 3.2, the mole fraction of alanine is:

$$x_{ala} = \frac{n_{ala}}{n_{ala}+n_{water}} = \frac{0.134\text{ mol}}{0.134\text{ mol} + (10^3\text{g}/18.02\text{ g/mol})}$$

$$x_{ala} = 2.41\times10^{-3}$$

3.21

$$X = \frac{n}{n+5.555} = 0.124$$

$$n\times(1-0.124) = 0.6888$$

$$n = 0.786$$

$$\text{mass} = 268.8\text{ g}$$

3.22 (a) The molar Gibbs energy of mixing can be calculated by using Equation 3.16:

$$\Delta G_m = RT(x_{N_2}\ln x_{N_2} + x_{O_2}\ln x_{O_2})$$

$$\Delta G_m = (8.31447\text{ J K}^{-1}\text{ mol}^{-1})\times(298.15\text{ K})\times(0.78\ \ln 0.78 + 0.22\ \ln 0.22)$$

$$\Delta G_m = -1.3\text{ kJ mol}^{-1}$$

(b) The molar entropy of mixing can be calculated at 298 K by using Equation 3.17b:

$$\Delta S_{\mathrm{m}} = -R(x_{\mathrm{N_2}} \ln x_{\mathrm{N_2}} + x_{\mathrm{O_2}} \ln x_{\mathrm{O_2}})$$

$$\Delta S_{\mathrm{m}} = -\frac{\Delta G_{\mathrm{m}}}{T} = \frac{1.3\ \mathrm{kJ\ mol^{-1}}}{298\ \mathrm{K}} = 4.4\ \mathrm{J\ K^{-1}}$$

The mixing is spontaneous.

3.23

$$\Delta G_{\mathrm{m}} = RT(x_1 \ln x_1 + x_2 \ln x_2)$$

$$\Delta G_{\mathrm{m}} = (8.31447\ \mathrm{J\ K^{-1}mol^{-1}}) \times (298.15\ \mathrm{K}) \times \left((0.78 \times \ln 0.78) + (0.22 \times \ln 0.22)\right)$$

$$\Delta G_{\mathrm{m}} = -1.3\ \mathrm{kJ\ mol^{-1}}$$

$$\Delta G_{\mathrm{m}} = RT(x_1 \ln x_1 + x_2 \ln x_2 + x_3 \ln x_3)$$

$$\Delta G_{\mathrm{m}} = (8.31447\ \mathrm{J\ K^{-1}mol^{-1}}) \times (298.15\ \mathrm{K}) \begin{pmatrix} (0.780 \times \ln 0.780) + (0.210 \times \ln 0.210) + \\ (0.0096 \times \ln 0.0096) \end{pmatrix}$$

$$\Delta G_{\mathrm{m}} = -1.40\ \mathrm{kJ\ mol^{-1}}$$

The additional change in molar Gibbs energy is $-0.1\ \mathrm{kJ\ mol^{-1}}$. The mixing is spontaneous.

The molar entropy can be calculated using the equation:

$$\Delta S_{\mathrm{m}} = -\frac{\Delta G_{\mathrm{m}}}{T} = -R\left(\sum_i x_i \ln x_i\right)$$

The additional change in molar entropy is 0 0.33 $\mathrm{J\ mol^{-1}}$.

3.24 The vapor pressure of seawater can be estimated by using the following equation:

$p_{\mathrm{seawater}} = x_{\mathrm{water}} p^*$.

Assuming the density of seawater is close to 1 g $\mathrm{mL^{-1}}$, we can calculate the mole fraction of water:

$$x_{\mathrm{water}} = \frac{n_{\mathrm{water}}}{n_{\mathrm{water}} + n_{\mathrm{solute}}} = \left(\frac{55.49\ \mathrm{mol}}{55.49\ \mathrm{mol} + 1.0\ \mathrm{mol}}\right) = 0.98$$

Where $(1000\ \mathrm{g/L\ H_2O})/(18.02\ \mathrm{mol\ L^{-1}\ H_2O}) = 55.49\ \mathrm{mol/L}$ was used.

The vapor pressure of seawater is then:

$$\mathrm{p_{seawater}} = \mathrm{x_{water} p^{O}} = 0.98 \times 2.338\ \mathrm{k\ Pa} = 2.29\ \mathrm{k\ Pa}$$

3.25 Hemoglobin could theoretically transport 20.1 mL of oxygen per 100 $\mathrm{cm^3}$ of blood. The volume of oxygen carried by 100 $\mathrm{cm^3}$ of blood changes from 19.5 (0.97×20.1 mL) to 15.07 mL (0.75×20.1 mL) as blood is flowing from the lungs in the capillary.

3.26 (a) The mass of nitrogen that could be dissolved in 100 g of water is calculated using Henry's law:

$$c = Kp = \left(0.18\times10^{-6}\ \mathrm{g_{N_2}\ g_{water}^{-1}\ atm^{-1}}\right)\times(4.0\ \mathrm{atm})\times0.7808$$

$$c = 0.56\ \mathrm{\mu g_{N_2}\ g_{water}^{-1}}$$

The mass of nitrogen that could be dissolved in 100 g of water at 4.0 atm is 56 μg. For 1.0 atm, this amount is 14 μg.

(b) The increase in nitrogen concentration in fatty tissue is by a factor of 16.

3.27 $$x_{CO_2} = \frac{55\ \mathrm{kPa}}{8.6\times10^{4}\,\mathrm{Torr}}\times\frac{1\ \mathrm{Torr}}{0.13332\ \mathrm{kPa}} = 4.797\times10^{-3}$$

3.28 The solubility of CO_2 in natural waters is:

(a)

$$[\mathrm{CO_2}] = K_{CO_2}p_{CO_2} = \left(3.39\times10^{-1}\ \mathrm{mol\ m^{-3}\ kPa^{-1}}\right)\times(4.0\ \mathrm{kPa})$$

$$[\mathrm{CO_2}] = 1.36\times10^{-3}\ \mathrm{mol\ L^{-1}}$$

(b)

$$[\mathrm{CO_2}] = K_{CO_2}p_{CO_2} = 3.39\times10^{-1}\mathrm{mol\ m^{-3}\ kPa^{-1}}\times(100.0\ \mathrm{kPa})$$

$$[\mathrm{CO_2}] = 33.9\times10^{-3}\mathrm{mol\ L^{-1}}$$

3.29

$$p = kx_{N_2}$$

$$x_{N_2} = \frac{0.78\times760\ \mathrm{Torr}}{6.51\times10^{7}\,\mathrm{Torr}} = 9.1\times10^{-6}$$

$$x_{N_2} = 9.1\times10^{-6} = \frac{n_{N_2}}{n_{N_2}+n_{H_2O}} = \frac{n_{N_2}}{n_{N_2}+55.55}$$

$$n_{N_2} = \frac{5.055\times10^{-4}}{(1-9.1\times10^{-6})} = 0.51\ \mathrm{mmol}$$

$$m_{N_2} = 0.51\ \mathrm{mmol\ kg^{-1}}$$

$$p = kx_{O_2}$$

$$x_{O_2} = \frac{0.21 \times 760\ \text{Torr}}{3.30 \times 10^7\ \text{Torr}} = 4.8 \times 10^{-6}$$

$$x_{O_2} = 4.8 \times 10^{-6} = \frac{n_{O_2}}{n_{O_2} + n_{H_2O}} = \frac{n_{O_2}}{n_{O_2} + 55.55}$$

$$n_{O_2} = \frac{2.686 \times 10^{-4}}{(1 - 4.8 \times 10^{-6})} = 0.27\ \text{mmol}$$

$$m_{O_2} = 0.27\ \text{mmol kg}^{-1}$$

3.30 The freezing point of water sweetened with sucrose can be calculated by using data from Table 3.4 and Equation 3.24 and assuming the density of water is 1 g mL^{-1}.

$$\Delta T_f = K_f b_B = \left(1.86\ \text{K kg mol}^{-1}\right) \times \frac{7.5\ \text{g}}{(342.30\ \text{g mol}^{-1}) \times\ (0.150\ \text{kg})}$$

$$\Delta T_f = 0.27\ \text{K}$$

The freezing point is estimated to be -0.27°C

3.31

	2A	$\rightleftarrows$	A_2
change	$-2x$		$+x$
[]	$c - 2x$		x

Total number of moles per liter: $c - x$

$$p = x_s p^* = \frac{n_s}{n_s + n_A + n_{A_2}} p^* = \frac{n_s p^*}{n_s + (c - x)}$$

$$\frac{1}{p} = \frac{n_s + (c - x)}{n_s p^*} = \frac{n_s + c - x}{n_s p^*}$$

$$\frac{n_s p^*}{p} - n_s - c = -x$$

$$\frac{n_s p^* - p n_s - pc}{p} = -x$$

$$\frac{pc - n_s\left(p^* - p\right)}{p} = x$$

$$K=\frac{[A_2]}{[A]^2}=\frac{x}{(c-2x)^2}=\frac{\dfrac{pc-n_s(p^*-p)}{p}}{\left\{c-\left[\dfrac{2pc-2n_s(p^*-p)}{p}\right]\right\}^2}$$

3.32 By using Equation 3.25b, the molarity of the solution of urea is found:

$$[\text{urea}]=\frac{\Pi}{RT}=\frac{120\text{ kPa}}{(8.31447\text{ L kPa K}^{-1}\text{mol}^{-1})\times(300\text{ K})}$$

$$[\text{urea}]=4.81\times10^{-2}\text{ mol L}^{-1}$$

Assuming ideal behavior,

$$\Delta T_f=K_f b_B=(1.86\text{ K kg mol}^{-1})\times(4.81\times10^{-2}\text{mol kg}^{-1})$$

$$\Delta T_f=0.09\text{ K}$$

The freezing point is estimated to be -0.09°C.

3.33 The relationship between h and mass concentration c can be found by substituting $\Pi=\rho gh$ into Equation 3.26 and using the definition [B] = c/M as shown below:

$$\Pi=\rho gh=[\text{B}]RT\{1+B[\text{B}]+..\}$$

$$h=\frac{[\text{B}]RT}{\rho g}\{1+B[\text{B}]+..\}$$

$$[\text{B}]=\frac{c}{M}$$

$$h=\frac{cRT}{M\rho g}\left\{1+B\frac{c}{M}+..\right\}$$

$$\frac{h}{c}=\frac{RT}{M\rho g}\left\{1+B\frac{c}{M}+..\right\}$$

$$\frac{h}{c}=\frac{RT}{M\rho g}+\left(\frac{BRT}{M^2\rho g}\right)c$$

Then, the molar mass of the enzyme can be obtained by plotting of h/c against mass concentration c and extrapolating to c =0.

The intercept of the graph is 1.784 $mg^{-1}cm^4$. The value of M is then:

$$M=\left(\frac{RT}{\rho g}\right)\times\frac{1}{h/c}$$

$$M=\frac{(8.31447\text{ J K}^{-1}\text{mol}^{-1})\times(293K)}{(1000\text{ kg m}^{-3})\times(9.81\text{ m s}^{-2})}\times\frac{1}{0.01784\text{ m}^4\text{ kg}^{-1}}$$

$$M=13.9\text{ kg mol}^{-1}$$

3.34 Using Equation 3.28, the osmotic virial coefficient could be calculated. The concentration of the polyanion will be taken as $9.1\times10^{-4}\text{mol L}^{-1}$.

$$B = \frac{v^2\left[\text{Cl}^-\right]}{4\left[\text{Cl}^{-1}\right]^2 + 2v\left[\text{Cl}^-\right]\left[\text{P}^{v-}\right]}$$

$$B = \frac{(20)^2 \times 0.02 \text{ mol L}^{-1}}{4 \times\left(0.02 \text{ mol L}^{-1}\right)^2 + (2\times 20)\times\left(0.02 \text{ mol L}^{-1}\right)\times(9.1\times10^{-4}\text{mol L}^{-1})}$$

$$B = 4.9 \times 10^3 \text{ mol L}^{-1}$$

Chapter 4
Chemical Equilibrium

4.7

$G6P(aq) + H_2O(l) \rightarrow G(aq) + P_i(aq)$, where G6P is glucose-6-phosphate, G is glucose, and P_i is inorganic phosphate.

$Gly(aq) + Ala(aq) \rightarrow Gly\text{-}Ala(aq) + H_2O(l)$

$Mg^{2+}(aq) + ATP^{4-}(aq) \rightarrow MgATP^{2-}(aq)$

$2\ CH_3COCOOH(aq) + 5\ O_2(g) \rightarrow 6\ CO_2(g) + 4\ H_2O(l)$

(a)

$$K = \frac{[G][P_i]}{[G6P]}$$

(b)

$$K = \frac{[Gly\text{-}Ala]}{[Gly][Ala]}$$

(c)

$$K = \frac{[MgATP^{2-}]}{[Mg^{2+}][ATP^{4-}]}$$

(d)

$$K = \frac{p_{CO_2}^6}{p_{O_2}^5[CH_3COCOOH]^2}$$

4.8 (a) The equilibrium constant will be $(3.4\times10^4)^{-1} = 2.9\times10^{-5}$.

(b) The equilibrium constant will be $(3.4\times10^4) = 1.1\times10^9$.

(c) The equilibrium constant will be $(3.4\times10^4)^{1/2} = 1.8\times10^2$.

4.9

$$\Delta_r G^{\ominus} = -RT \ln K$$

$$\Delta_r G^{\ominus} = -(8.31447 \text{ J K}^{-1}\text{mol}^{-1}) \times (310\text{K}) \times (\ln 8.1 \times 10^2)$$

$$\Delta_r G^{\ominus} = 17.3 \text{ J mol}^{-1}$$

4.10 The relationship between equilibrium constant and the standard Gibbs energy is given by Equation 4.8.

$$\frac{K_1}{K_2} = 10 = e^{-\left(\frac{\Delta_r G_1^{\ominus}}{RT} - \frac{\Delta_r G_2^{\ominus}}{RT}\right)} = e^{-\frac{1}{RT}\left(\Delta_r G_1^{\ominus} - \Delta_r G_2^{\ominus}\right)} = e^{-\left[\frac{1}{(8.31447 \text{ J K}^{-1} \text{ mol}^{-1}) \times (300 \text{ K})} \times \left((-300\times10^3 \text{ kJ mol}^{-1}) - \Delta_r G_2^{\ominus}\right)\right]}$$

$$\ln 10 = -\left[\frac{1}{(8.31447 \text{ J K}^{-1} \text{ mol}^{-1}) \times (300 \text{ K})} \times \left((-300 \times 10^3 \text{ J mol}^{-1}) - \Delta_r G_2^{\ominus}\right)\right]$$

$$\Delta_r G_2^{\ominus} = \left[(\ln 10) \times \left(8.31447 \text{ J K}^{-1} \text{ mol}^{-1}\right) \times (300 \text{ K})\right] - (300 \times 10^3 \text{ J mol}^{-1})$$

$$\Delta_r G_2^{\ominus} = -294 \times 10^3 \text{ J mol}^{-1}$$

4.11 $\Delta_r G^{\ominus} = -RT \ln K$

If K has a value of 1 $\Delta_r G^{\ominus}$ will be zero.

4.12 Using Equation 4.8, the equilibrium constants for glucose-1-phosphate, glucose-6-phosphate and glucose-3-phosphate are (respectively):

$$K = e^{-\Delta_r G^{\ominus}/RT} = e^{\frac{21\times10^3 \text{J mol}^{-1}}{(8.31447 \text{ J K}^{-1} \text{ mol}^{-1})\times(310 \text{ K})}} = 3.5 \times 10^3$$

$$K = e^{-\Delta_r G^{\ominus}/RT} = e^{\frac{14\times10^3 \text{J mol}^{-1}}{(8.31447 \text{ J K}^{-1} \text{ mol}^{-1})\times(310 \text{ K})}} = 2.3 \times 10^2$$

$$K = e^{-\Delta_r G^{\ominus}/RT} = e^{\frac{9.2\times10^3 \text{J mol}^{-1}}{(8.31447 \text{ J K}^{-1} \text{ mol}^{-1})\times(310 \text{ K})}} = 35$$

4.13 ATP(aq) + H_2O(l) → ADP(aq) + P_i(aq)

(a)

$$\Delta_r G = \Delta_r G^{\ominus} + RT \ln Q$$

$$\Delta_r G = (-31 \times 10^3 \text{ J mol}^{-1}) + (8.31447 \text{ J K}^{-1}\text{mol}^{-1}) \times (310.15 \text{ K}) \ln \frac{(1\times10^{-3}) \times (1\times10^{-3})}{1\times10^{-3}}$$

$$\Delta_r G = -48 \text{ k J mol}^{-1}$$

(b)

$$\Delta_r G = \Delta_r G^{\ominus} + RT \ln Q$$

$$\Delta_r G = (-31\times10^3 \text{ J mol}^{-1}) + (8.31447 \text{ J K}^{-1}\text{mol}^{-1})\times(310.15 \text{ K})\times\ln\frac{(1\times10^{-6})\times(1\times10^{-6})}{1\times10^{-6}}$$

$$\Delta_r G = -67 \text{ kJ mol}^{-1}$$

4.14 Assuming the concentration of Na^+ ions is equal to the activity of Na^+ ions, the chemical potential of Na^+ can be expressed using Equation 4.3:

$$\mu_{\text{Na}^+} = \mu^{\ominus}_{\text{Na}^+} + RT \ln [\text{Na}^+]$$

The Gibbs energy difference is then:

$$\Delta_r G = \mu_{\text{inside}} - \mu_{\text{outside}} = RT \ln [\text{Na}^+]_{\text{inside}} - RT \ln [\text{Na}^+]_{\text{outside}}$$

$$\Delta_r G = RT \ln \frac{[\text{Na}^+]_{\text{inside}}}{[\text{Na}^+]_{\text{outside}}} = (8.31447 \text{ J K}^{-1} \text{ mol}^{-1})\times(310 \text{ K})\times(\ln \frac{10}{140})$$

$$\Delta_r G = -6.8 \text{ kJ mol}^{-1}$$

4.15 $\Delta_r G^{\ominus} = \Delta_r H^{\ominus} - T\Delta S^{\ominus}$

Since $\Delta_r H^{\ominus}$ is negative and $\Delta S^{\ominus}$ is positive, K is > than 1 at all temperatures.

4.16 (a) The stability of a double helix is related to the number of hydrogen bonds. Increasing the number of base pairs should contribute to the stability of the double helix.

(b)

$$\frac{K_1}{K_2} = \frac{2.0\times10^5}{5.0\times10^3} = e^{-\left(\frac{\Delta_r G_1^{\ominus}}{RT} - \frac{\Delta_r G_2^{\ominus}}{RT}\right)} = e^{-\frac{1}{RT}\left(\Delta_r G_1^{\ominus} - \Delta_r G_2^{\ominus}\right)}$$

$$\ln 40 = -\left[\frac{1}{(8.31447 \text{ J K}^{-1} \text{ mol}^{-1})\times(300 \text{ K})} \times \left(\Delta_r G_1^{\ominus} - \Delta_r G_2^{\ominus}\right)\right]$$

$$\Delta_r G_2^{\ominus} - \Delta_r G_1^{\ominus} = (\ln 40)\times\left((8.31447 \text{ J K}^{-1} \text{ mol}^{-1})\times 300 \text{ K}\right)$$

$$\Delta_r G_2^{\ominus} - \Delta_r G_1^{\ominus} = 9.2 \text{ kJ mol}^{-1}$$

4.17 (a) The complete combustion of glucose gives:

$$C_6H_{12}O_6(s)+6O_2(g) \rightarrow 6CO_2(g)+6H_2O(l) \qquad \Delta_r G^{\ominus} = -2880 \text{ kJ mol}^{-1}$$

The synthesis of ATP from ADP and HPO_4^{2-} requires 31 kJ mol^{-1} as follows:

$$ADP^{3-}(aq)+HPO_4^{2-}(aq)+H_3O^+(aq) \longrightarrow ATP^{4-}(aq)+H_2O(l)$$

The percent efficiency of aerobic respiration under biochemical standard conditions is:

$$\text{efficiency} = \frac{38 \times 31 \text{ kJ}}{2880 \text{ kJ}} \times 100\% = 41\%$$

(b) The Gibbs energy of the reaction for the combustion of glucose under typical conditions found in living cells is:

$$\Delta_r G = \Delta_r G^{\ominus} + RT \ln Q$$

$$\Delta_r G = (-2880 \text{ kJ mol}^{-1}) + (8.31447 \times 10^{-3} \text{ kJ K}^{-1}\text{mol}^{-1}) \times (310 \text{ K}) \times \ln \frac{(5.3 \times 10^{-2})^6}{(0.132)^6}$$

$$\Delta_r G = -2894 \text{ kJ mol}^{-1}$$

The Gibbs energy of the reaction for the synthesis of ATP from ADP and HPO_4^{2-} under typical conditions found in living cells is:

$$\Delta_r G = \Delta_r G^{\oplus} + RT \ln Q^{\oplus}$$

$$\Delta_r G = (+31 \text{ kJ mol}^{-1}) + (8.31447 \times 10^{-3} \text{ kJ K}^{-1}\text{mol}^{-1}) \times (310 \text{ K})$$

$$\times \ln \frac{1.0 \times 10^{-4}}{(1.0 \times 10^{-4})(1.0 \times 10^{-4})(3.98 \times 10^{-8} / 1 \times 10^{-7})}$$

$$\Delta_r G = +57 \text{ kJ mol}^{-1}$$

The percent efficiency of aerobic respiration under these conditions is:

$$\text{efficiency} = \frac{38 \times 57 \text{ kJ}}{2894 \text{ kJ}} \times 100\% = 75\%$$

4.18

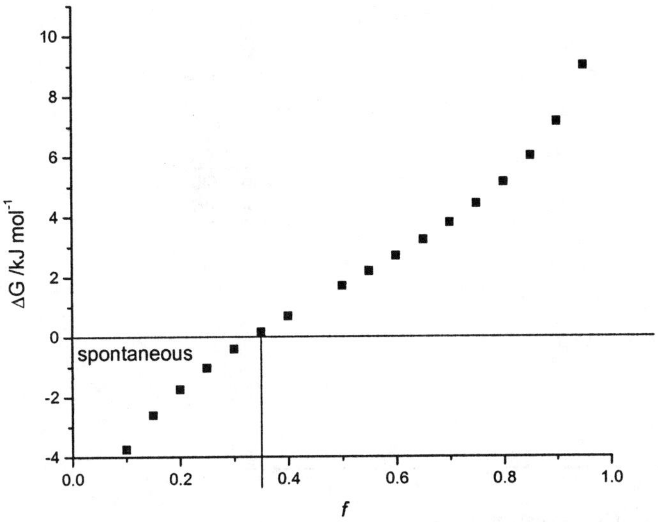

The equations below show the relationship between f, Q and $\Delta_r G^\circ$.

$$f = \frac{Q}{Q+1} \longrightarrow Q = \frac{1}{\frac{1}{f}-1} = \frac{f}{1-f}$$

$$\Delta_r G = \Delta_r G^\circ + RT \ln\left(\frac{f}{1-f}\right)$$

4.19 (a) The relation between s and p is given by:

$$\log\frac{s}{1-s} = \nu \log p - \nu \log K$$

$s = 0.5,\ p = K$

From the graph we can only estimate the values of K. Fractional saturation of Mb and Hb as a function of p are obtained using the following values of K: 26 Torr for Hb, and 2.8 Torr for Mb.

Fractional saturation of Mb and Hb for different p values are listed below:

p/Torr	7.5	11.2	18.7	30.0	60.0
s (Hb)	0.02987	0.08744	0.28598	0.59893	0.91228
s (Mb)	0.72818	0.80073	0.87008	0.91464	0.95542

A graph of s against oxygen partial pressure is shown below:

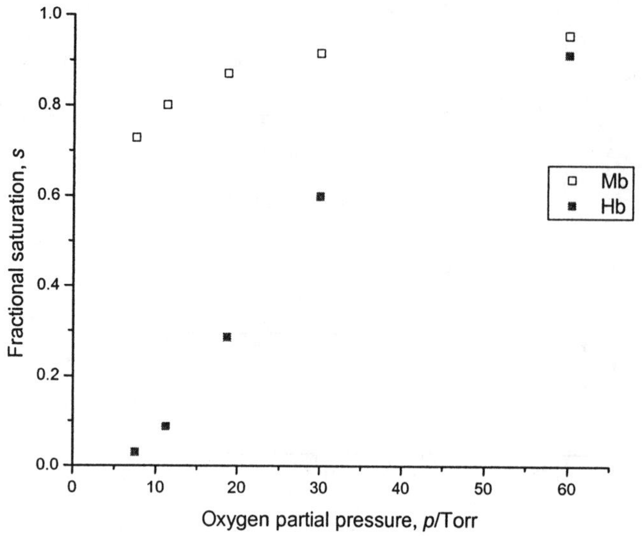

(b) Values of s for Hb and Mb assuming ν is equal to 4 are listed in the table below:

p/Torr	7.5	11.2	18.7	30.0	60.0
s (Hb)	0.00688	0.03389	0.21299	0.63946	0.96596
s (Mb)	0.98095	0.99618	0.9995	0.99992	1

The graph below compares fractional saturation of Hb and Mb against p using two different values of ν.

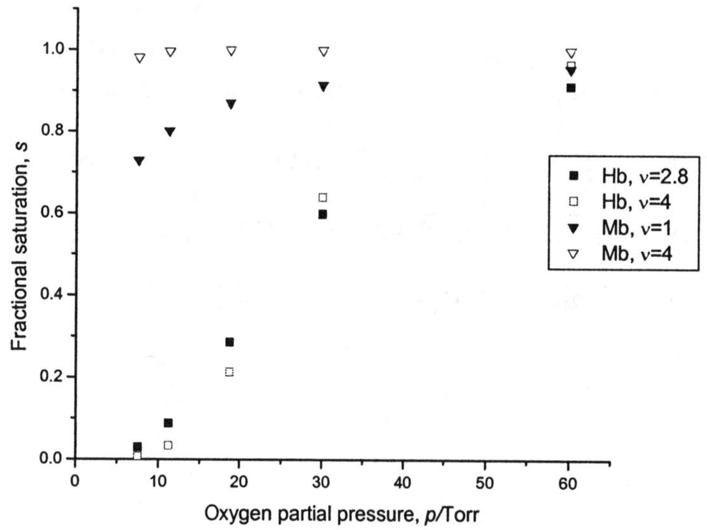

4.20 Glucose, urea and ethanol are classified as exergonic because they have negative $\Delta_f G^{\circ}$. Octane is classified as endergonic because its $\Delta_f G^{\circ}$ is positive.

4.21 (a)

$$\Delta_r H^{\circ} = \left[\left(-12\times 393.51 \text{ kJ mol}^{-1}\right)+\left(-11\times 285.83 \text{ kJ mol}^{-1}\right)\right]-\left[\left(-2222 \text{ kJ mol}^{-1}\right)\right]$$

$$\Delta_r H^{\circ} = -5644 \text{ kJ mol}^{-1}$$

$$\Delta_r S^{\circ} = \left[\left(12\times 213.74 \text{ J K}^{-1}\text{mol}^{-1}\right)+\left(11\times 69.91 \text{ J K}^{-1}\text{mol}^{-1}\right)\right] - \left[\left(360.2 \text{ J K}^{-1}\text{mol}^{-1}\right)+\left(12\times 205.138 \text{ J K}^{-1}\text{mol}^{-1}\right)\right]$$

$$\Delta_r S^{\circ} = +0.512 \text{ kJ K}^{-1}\text{mol}^{-1}$$

$$\Delta_r G^{\circ} = \Delta_r H^{\circ} - T\Delta_r S^{\circ}$$

$$\Delta_r G^{\circ} = (-5644 \text{ kJ mol}^{-1}) - (298 \text{ K})\times(0.512 \text{ kJ K}^{-1}\text{mol}^{-1})$$

$$\Delta_r G^{\circ} = -5796 \text{ kJ mol}^{-1}$$

(b) i. The maximum energy that can be extracted as heat is $q = (5644 \text{ kJ mol}^{-1})\times(2.9 \text{ mol}) = 1.6\times 10^4 \text{ kJ}$. In this case, the term $T\Delta S$ is the amount of heat obtained from surroundings.

ii. The maximum non-expansion work that can be extracted per kg of sucrose is $w = (5796 \text{ kJ mol}^{-1})\times(2.9 \text{ mol}) = 1.68\times 10^4 \text{ kJ}$

4.22 $C_6H_{12}O_6(s) + 6\ O_2(g) \rightarrow 6\ CO_2(g) + 6\ H_2O(l)$

$w = -p_{ex}\Delta V = -p_{ex}(V_f - V_i) = -p_{ex}(V_{CO_2} - V_{O_2})$

Work of expansion is zero because final and initial gaseous volumes are the same. The change in Gibbs energy can be identified with the maximum non-expansion work.

The change in Gibbs energy for the combustion of glucose is:

$$\Delta_r G^{\circ} = \left[\left(-6\times 394.36 \text{ kJ mol}^{-1}\right)+\left(-6\times 237.13 \text{ kJ mol}^{-1}\right)\right]-\left[\left(-910 \text{ kJ mol}^{-1}\right)\right]$$

$$\Delta_r G^{\circ} = -2878 \text{ kJ mol}^{-1}$$

The maximum non-expansion work that can be extracted from 1 kg of glucose is:

$$w = 2878 \text{ kJ mol}^{-1} \times \frac{1000 \text{ g}}{180.16 \text{ g mol}^{-1}} = 1.59\times 10^4 \text{ kJ}$$

It is more energy effective to ingest sucrose than glucose.

4.23 (a)

$$\text{pyruvate}(aq) \longrightarrow \text{ethanal}(g) + CO_2(g)$$

$$\Delta_r G^{-} = -394.36 \text{ kJ mol}^{-1} - 133 \text{ kJ mol}^{-1} + 474 \text{ kJ mol}^{-1}$$

$$\Delta_r G^{-} = -53.4 \text{ kJ mol}^{-1}$$

(b) Less negative

4.24 The standard biological Gibbs energy can be calculated as shown below:

$$\Delta_r G^{-} = \Delta_r G^{-'} + RT \ln \frac{1}{1\times 10^{-7}}$$

$$\Delta_r G^{-} = \Delta_r G^{-'} + (7 \times RT \ \ln \ 10)$$

$$\Delta_r G^{-} = \Delta_r G^{-'} + 41.54 \text{ kJ mol}^{-1}$$

$$\Delta_r G^{-} = -25.0 \text{ kJ mol}^{-1}$$

4.25 AMP + $H_2O \rightarrow P_i + H^+$

$$\Delta_r G^{-} = \Delta_r G^{-'} + RT \ln \frac{1}{1\times 10^{-7}}$$

$$\Delta_r G^{-} = \Delta_r G^{-'} + 7 \times RT \ln 10$$

$$\Delta_r G^{-} = \Delta_r G^{-'} + 39.93 \text{ kJ mol}^{-1}$$

$$\Delta_r G^{-} = 26 \text{ kJ mol}^{-1}$$

4.26 We can estimate the biological standard Gibbs energies by using transfer potentials from Table 4.3

(a) The biological standard Gibbs energy for GTP(aq) + ADP(aq) → GDP(aq) + ATP(aq) is equal to zero, as discussed in Section 4.8.

(b) In the reaction: Glycerol(aq) + ATP(aq) → glycerol-1-phosphate + ADP(aq) The phosphate group transfer potentials for ADP and ATP are the same and we only have the group transfer potentials for glycerol-1-phosphate which is −10 kJ mol^{-1}.

(c) For the reaction: 3-Phosphoglycerate(aq) + ATP(aq) → 1,3-bis(phospho)glycerate(aq) + ADP(aq), we only have the group transfer potentials for 1,3-bis(phospho)glycerate, which is −49 kJ mol^{-1}.

4.27

$$K = e^{-\Delta_r G^{-}/RT}$$

$$\ln K = -\frac{\Delta_r G^{-}}{RT} = -\frac{\left(\Delta_r H^{-} - T\Delta_r S^{-}\right)}{RT} = \frac{-\Delta_r H^{-}}{RT} + \frac{\Delta_r S^{-}}{R}$$

A plot of ln$K$ against $1/T$ allows the determination of the standard reaction enthalpy, since the slope of such a graph is $\dfrac{-\Delta_r H^{-}}{R}$.

4.28 The dependence of K with temperature is given by:

$$\ln\ K = -\frac{\Delta_r H^{-}}{RT} + \frac{\Delta_r S^{-}}{R}$$

A graph of ln K versus temperature is shown below:

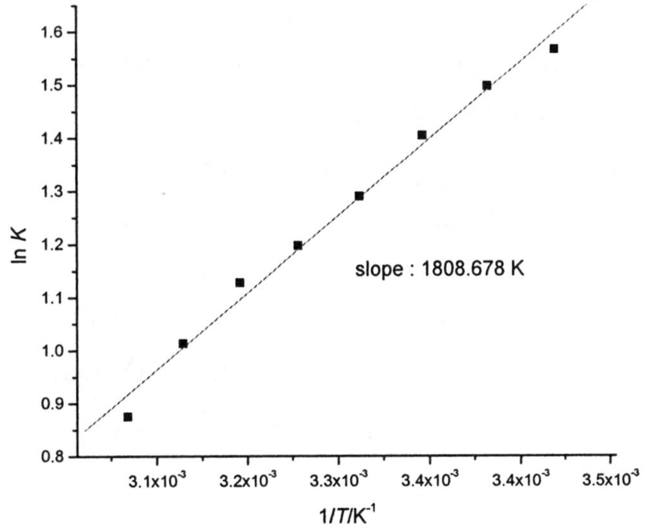

$$-\frac{\Delta_r H^{-}}{R} = \text{slope} = \ 1808.678\ \text{K}$$

$$\Delta_r H^{-}\ = -15.04\ \text{kJ mol}^{-1}$$

4.29

$$\ln K^{'} - \ln K = \frac{\Delta H}{R}\left(\frac{1}{T} - \frac{1}{T^{'}}\right)$$

$$\ln \frac{K^{'}}{K} = \ln 2 = 0.693 = \frac{\Delta H}{8.31447 \text{ J K}^{-1}\text{mol}^{-1}}\left(\frac{1}{298 \text{ K}} - \frac{1}{308 \text{ K}}\right)$$

$$\Delta H = 52.9 \text{ kJ}$$

$$\ln \frac{K^{'}}{K} = \ln 0.5 = -0.693 = \frac{\Delta H}{8.31447 \text{ J K}^{-1}\text{mol}^{-1}}\left(\frac{1}{298 \text{ K}} - \frac{1}{308 \text{ K}}\right)$$

$$\Delta H = -52.9 \text{ kJ}$$

4.30 According to Bronsted-Lowry theory:

(a) $H_2PO_4^- \text{ (aq)} + H_2O(l) \rightarrow H_3O^+(aq) + HPO_4^{2-} \text{ (aq)}$

$HPO_4^{2-} \text{ (aq)} + H_2O(l) \rightarrow H_3O^+(aq) + PO_4^{3-} \text{ (aq)}$

(b) $CH_3CHOHCOOH \text{ (aq)} + H_2O(l) \rightarrow H_3O^+(aq) + CH_3CHOHCOO^- \text{ (aq)}$

(c) $HOOCCH_2CH_2CH(NH_2)COOH \text{ (aq)} + H_2O(l) \rightarrow$

$H_3O^+(aq) + HOOCCH_2CH_2CH(NH_2)COO^- \text{ (aq)}$

$HOOCCH_2CH_2CH(NH_2)COO^- \text{ (aq)} + H_2O(l) \rightarrow$

$H_3O^+(aq) + {}^-OOCCH_2CH_2CH(NH_2)COO^- \text{ (aq)}$

(d) $NH_2CH_2COOH \text{ (aq)} + H_2O(l) \rightarrow H_2O(l) + {}^+NH_3CH_2COO^- \text{ (aq)}$

(e) $HOOCCOOH \text{ (aq)} + H_2O(l) \rightarrow H_3O^+(aq) + HOOCCOO^- \text{ (aq)}$

$HOOCCOO^- \text{ (aq)} + H_2O(l) \rightarrow H_3O^+(aq) + {}^-OOCCOO^- \text{ (aq)}$

4.31 (a)

$K_w = 2.5 \times 10^{-14} = x^2$

$x = [H_3O^+] = \sqrt{2.5 \times 10^{-14}} = 1.6 \times 10^{-7} \text{ M}$

$pH = 6.8$

(b)

$K_w = [H_3O^+][OH^-] = 2.5 \times 10^{-14}$

$$[OH^-] = \frac{2.5 \times 10^{-14}}{1.6 \times 10^{-7}} = 1.6 \times 10^{-7}$$

$pOH = 6.8$

4.32 (a) The chemical equation for the autodeuterolysis of D_2O is:

$D_2O + D_2O \rightarrow D_3O^+ + OD^-$

(b) $pK_W = 14.9$

(c) The molar concentrations of D_3O^+ and OD^- are:

$K_W = [D_3O^+][OD^-] = 1.35\times10^{-15}$

$[D_3O^+] = [OD^-] = \sqrt{1.35\times10^{-15}} = 3.67\times10^{-8}$ M

(d) pD = pOD = 7.43

(e) pD + pOD = pK_W

4.33 $pH = -\log[H_3O^+]$ and $pOH = 14 - pH$

(a) pH = 4.8, pOH = 9.2

(b) pH = 2.8, pOH = 11

(c) pH = pOH = 7

(d) pH = 4.30, pOH = 9.69

4.34 (a) The net ionic equation between HCl and NaOH is:

$H_3O^+(aq) + OH^-(aq) \rightarrow 2\ H_2O(l)$

We need to calculate the amount (moles) of H_3O^+ and OH^-

Moles $H_3O^+ = 3.60\times10^{-3}$

Moles $OH^- = 3.12\times10^{-3}$

The pH is determined by the slight excess of $[H_3O^+]$=

$(0.475\times10^{-3}\ \text{mol})/(0.0500\text{L})$

$[H_3O^+] = 9.5\times10^{-3}$ mol L^{-1}

pH = 2.02

(b) This case is the same as described in part (a).

Moles $H_3O^+ = 3.75\times10^{-3}$

Moles $OH^- = 5.25\times10^{-3}$

The pH is determined by the slight excess of $[OH^-] = (1.5\times10^{-3}\ \text{mol})/(0.060\ \text{L})$

$[OH^-]$= 0.025 mol L^{-1}, $[H_3O^+] = 4\times10^{-13}$ mol L^{-1}

pH = 12.4

(c) This case is the same as described in part (a).

Moles $H_3O^+ = 4.66\times10^{-3}$

Moles $OH^- = 3.00\times10^{-3}$

The pH is determined by the slight excess of $[H_3O^+]$

$(1.66\times10^{-3}\ \text{mol})/(0.0312\ \text{L})$

$[H_3O^+] = 5.32\times10^{-2}$ mol L^{-1}

pH = 1.27

4.35 (a) pH < 7

$NH_4^+(aq) + H_2O(l) \rightleftharpoons NH_3(aq) + H_3O^+(aq)$

(b) pH > 7

$CO_3^{2-}(aq) + H_2O(l) \rightleftharpoons HCO_3^-(aq) + OH^-(aq)$

(c) pH >7

$F^-(aq) + H_2O(l) \rightleftharpoons HF(aq) + OH^-(aq)$

(d) pH = 7

4.36 The pH of a salt can be solved following the procedure described in Example 4.5.
(a) The concentration of KCH_3CO_2 is 0.34 M. This salt consists of a neutral ion K^+, and a base $CH_3CO_2^-$.

$CH_3CO_2^-(aq) + H_2O(l) \rightarrow CH_3CO_2H(aq) + OH^-(aq)$

	$CH_3CO_2^-$	CH_3CO_2H	OH^-
Initial concentration/(mol L^{-1})	0.34	0	0
Change to reach equilibrium/(mol L^{-1})		$+x$	$+x$
Equilibrium concentration/(mol L^{-1})	$0.34 - x$	x	x

$$K_b = \frac{x^2}{0.34 - x} = 5.6\times10^{-10}$$

$x = 1.38\times10^{-5}$ M = $[OH^-]$

pH = 9.13

(b) The concentration of NH_4Br is 0.38 M. This salt consists of a neutral ion Br^-, and an acid NH_4^+

$NH_4^+(aq) + H_2O(l) \rightarrow NH_3(aq) + H_3O^+(aq)$

	NH_4^+	NH_3	H_3O^+
Initial concentration / (mol L^{-1})	0.38	0	0
Change to reach equilibrium / (mol L^{-1})	$-x$	$+x$	$+x$
Equilibrium concentration / (mol L^{-1})	$0.38 - x$	x	x

$$K_a = \frac{x^2}{0.38 - x} = 5.6\times10^{-10}$$

$x = 1.46\times10^{-5}$ M = $[H_3O^+]$

pH = 4.83

(c) 0% because Br^- is a neutral ion.

4.37 (a)

$pH = pK_a = 3.08$

$K_a = 10^{-3.08} = 8.31 \times 10^{-4}$

(b)

$$pH = pK_a - \log \frac{[acid]}{[base]}$$

$pH = 3.08 - \log 2$

$pH = 2.78$

4.38 (a) $CH_3CH(OH)COOH(aq) + H_2O(l) \rightarrow CH_3CH(OH)COO^-(aq) + H_3O^+(aq)$

	Lactic acid	Lactate ion	H_3O^+
Initial concentration / (mol L^{-1})	0.120	0	0
Change to reach equilibrium / (mol L^{-1})	$-x$	$+x$	$+x$
Equilibrium concentration / (mol L^{-1})	$0.120 - x$	x	x

$$K_a = \frac{x^2}{0.12 - x} = 8.4 \times 10^{-4}$$

$x^2 + 8.4 \times 10^{-4} x - 1.00 \times 10^{-4} = 0$

$[H_3O^+] = 9.58 \times 10^{-3}$ M

$pH = 2.02$

$pOH = 11.9$

fraction = 0.08

(b) $CH_3CH(OH)COOH(aq) + H_2O(l) \rightarrow CH_3CH(OH)COO^-(aq) + H_3O^+(aq)$

	Lactic acid	Lactate ion	H_3O^+
Initial concentration /(mol L^{-1})	1.4×10^{-4}	0	0
Change to reach equilibrium /(mol L¯1)	$-x$	$+x$	$+x$
Equilibrium concentration /(mol L^{-1}) Equilibrium concentration/(mol L^{-1})	$1.4 \times 10^{-4} - x$	x	x

$$K_a = \frac{x^2}{1.4 \times 10^{-4} - x} = 8.4 \times 10^{-4}$$

$x^2 + 8.4 \times 10^{-4} x - 1.18 \times 10^{-7} = 0$

$[H_3O^+] = 1.22 \times 10^{-4}$ M

$pH = 3.91$

$pOH = 10.9$

fraction = 0.87

(c) $NH_4^+(aq) + H_2O(l) \rightarrow NH_3(aq) + H_3O^+(aq)$

	NH_4^+	NH_3	H_3O^+
Initial concentration /(mol L^{-1})	0.15	0	0
Change to reach equilibrium /(mol L^{-1})	$-x$	$+x$	$+x$
Equilibrium concentration /(mol L^{-1})	$0.15 - x$	x	x

$$K_a = \frac{x^2}{0.15 - x} = 5.6\times10^{-10}$$

$x = 9.16\times10^{-6}\ \text{M} = [H_3O^+]$

pH = 5.04

pOH = 8.96

fraction = 6.1×10^{-5}

(d) $CH_3CO_2^-(aq) + H_2O(l) \rightarrow CH_3CO_2H(aq) + OH^-(aq)$

	$CH_3CO_2^-$	CH_3CO_2H	OH^-
Initial concentration /(mol L^{-1})	0.15	0	0
Change to reach equilibrium /(mol L^{-1})	$-x$	$+x$	$+x$
Equilibrium concentration /(mol L^{-1})	$0.15 - x$	x	x

$$K_b = \frac{x^2}{0.15 - x} = 5.6\times10^{-10}$$

$x = 9.16\times10^{-6}\ \text{M} = [OH^-]$

pH = 8.96

pOH = 5.04

fraction = 6.1×10^{-5}

(e) $(CH_3)N(aq) + H_2O(l) \rightarrow (CH_3)NH^+(aq) + OH^-(aq)$

	$(CH_3)N$	$(CH_3)NH^+$	OH^-
Initial concentration /(mol L^{-1})	0.112	0	0
Change to reach equilibrium /(mol L^{-1})	$-x$	$+x$	$+x$
Equilibrium concentration /(mol L^{-1})	$0.112 - x$	x	x

$$K_b = \frac{x^2}{0.112 - x} = 6.5\times10^{-5}$$

x = $2.69\times10^{-3}\ \text{M} = [OH^-]$

pH = 11.4

pOH = 2.57

fraction = 2.4×10^{-2}

4.39

$$H_2Gly^+(aq) + H_2O(l) \rightleftharpoons HGly(aq) + H_3O^+(aq)$$

$$K_{a1} = \frac{[HGly][H_3O^+]}{[H_2Gly^+]} = \frac{[HGly]H}{[H_2Gly^+]}$$

$$H = [H_3O^+]$$

$$HGly(aq) + H_2O(l) \rightleftharpoons Gly^-(aq) + H_3O^+(aq)$$

$$K_{a2} = \frac{[Gly^-]H}{[HGly]}$$

$$G = \text{total}[\,] = [H_2Gly^+] + [HGly] + [Gly^-]$$

$$[Gly^-] = \frac{K_{a2}[HGly]}{H} = \frac{K_{a2}K_{a1}[H_2Gly^+]}{H^2}$$

$$G = [H_2Gly^+] + \frac{K_{a1}[H_2Gly^+]}{H} + \frac{K_{a2}K_{a1}[H_2Gly^+]}{H^2}$$

$$f_{(H_2Gly^+)} = \frac{[H_2Gly^+]}{G} = \frac{1}{1 + \frac{K_{a1}}{H} + \frac{K_{a2}K_{a1}}{H^2}} = \frac{H^2}{H^2 + K_{a1}H + K_{a2}K_{a1}} = \frac{H^2}{K}$$

$$K = H^2 + K_{a1}H + K_{a2}K_{a1}$$

$$f_{(HGly)} = \frac{[HGly]}{G} = \frac{\frac{K_{a1}[H_2Gly^+]}{H}}{[H_2Gly^+] + \frac{K_{a1}[H_2Gly^+]}{H} + \frac{K_{a2}K_{a1}[H_2Gly^+]}{H^2}}$$

$$f_{(HGly)} = \frac{HK_{a1}}{K}$$

$$f_{(Gly^-)} = \frac{[Gly^-]}{G} = \frac{\frac{K_{a2}K_{a1}[H_2Gly^+]}{H^2}}{[H_2Gly^+] + \frac{K_{a1}[H_2Gly^+]}{H} + \frac{K_{a2}K_{a1}[H_2Gly^+]}{H^2}} =$$

$$f_{(Gly^-)} = \frac{K_{a2}K_{a1}}{H^2 + K_{a1}H + K_{a2}K_{a1}} = \frac{K_{a2}K_{a1}}{K}$$

The fractional composition of the protonated and deprotonated forms of glycine in an aqueous solution are plotted as a function of pH:

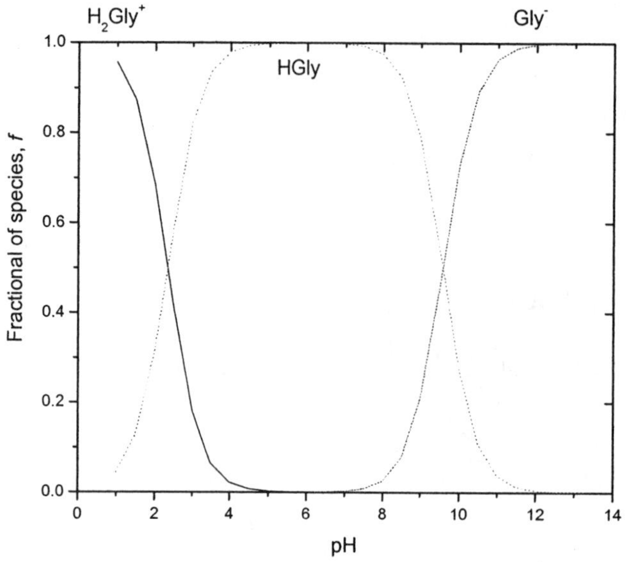

4.40 This case is very similar to Case Study 4.3.

$$H_3Tyr^+(aq) + H_2O(l) \rightleftharpoons H_2Tyr(aq) + H_3O^+(aq)$$

$$K_{a1} = \frac{[H_2Tyr][H_3O^+]}{[H_3Tyr^+]} = \frac{[H_2Tyr]H}{[H_3Tyr^+]}$$

$$H = [H_3O^+]$$

$$H_2Tyr(aq) + H_2O(l) \rightleftharpoons HTyr^-(aq) + H_3O^+(aq)$$

$$K_{a2} = \frac{[HTyr^-]H}{[H_2Tyr]}$$

$$HTyr^-(aq) + H_2O(l) \rightleftharpoons Tyr^{2-}(aq) + H_3O^+(aq)$$

$$K_{a3} = \frac{[Tyr^{2-}]H}{[HTyr^-]}$$

$$T = [H_3Tyr^+] + [H_2Tyr] + [HTyr^-] + [Tyr^{2-}]$$

$$[Tyr^{2-}] = \frac{K_{a3}[HTyr^-]}{H} = \frac{K_{a3}K_{a2}[H_2Tyr]}{H^2} = \frac{K_{a3}K_{a2}K_{a1}[H_3Tyr^+]}{H^3}$$

$$[HTyr^-] = \frac{K_{a2}K_{a1}[H_3Tyr^+]}{H^2}$$

$$[H_2Tyr] = \frac{K_{a1}[H_3Tyr^+]}{H}$$

$$T = [\mathrm{H_3Tyr^+}] + \frac{K_{a1}[\mathrm{H_3Tyr^+}]}{H} + \frac{K_{a2}K_{a1}[\mathrm{H_3Tyr^+}]}{H^2} + \frac{K_{a3}K_{a2}K_{a1}[\mathrm{H_3Tyr^+}]}{H^3}$$

$$T = [\mathrm{H_3Tyr^+}] \times \frac{\left(H^3 + H^2K_{a1} + HK_{a2}K_{a1} + K_{a3}K_{a2}K_{a1}\right)}{H^3}$$

$$K = H^3 + H^2K_{a1} + HK_{a2}K_{a1} + K_{a3}K_{a2}K_{a1}$$

$$T = [\mathrm{H_3Tyr^+}] \times \frac{K}{H^3}$$

$$f_{(\mathrm{H_3Tyr^+})} = \frac{[\mathrm{H_3Tyr^+}]}{T} = \frac{H^3}{K}$$

$$f_{(\mathrm{H_2Tyr})} = \frac{[\mathrm{H_2Tyr}]}{T} = \frac{H^2K_{a1}}{K}$$

$$f_{(\mathrm{HTyr^-})} = \frac{[\mathrm{HTyr^-}]}{T} = \frac{HK_{a2}K_{a1}}{K}$$

$$f_{(\mathrm{Tyr^{2-}})} = \frac{[\mathrm{Tyr^{2-}}]}{T} = \frac{K_{a3}K_{a2}K_{a1}}{K}$$

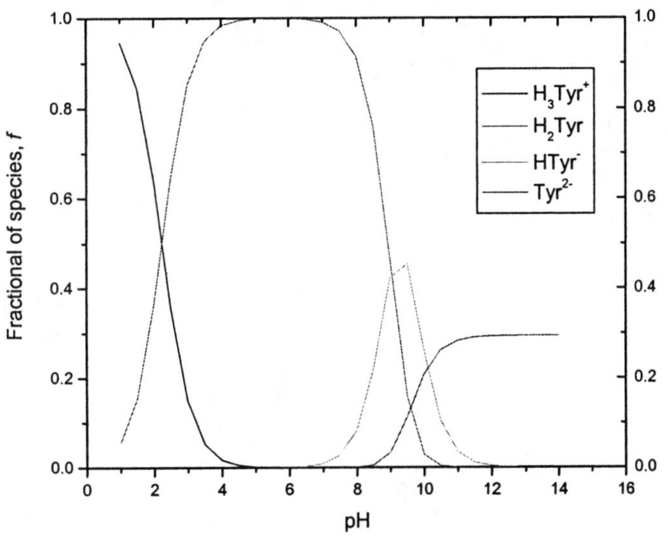

4.41 (a) $H_3BO_3(aq) + H_2O(l) \rightleftharpoons H_2BO_3^-(aq) + H_3O^+(aq)$

$1.0\times10^{-4} - x$	x	x

$$K_a = \frac{x^2}{1.0\times10^{-4} - x} = 7.2\times10^{-10}$$

$$7.2\times10^{-10} = \frac{x^2}{1.0\times10^{-4}}$$

$x = [H_3O^+] = 2.7\times10^{-7}$ M

pH = 6.6

(b) $H_3PO_4(aq) + H_2O(l) \rightleftharpoons H_2PO_4^-(aq) + H_3O^+(aq)$

$0.015 - x$	x	x

$$K_a = \frac{x^2}{0.015 - x} = 7.6\times10^{-3}$$

$$7.6\times10^{-3} = \frac{x^2}{0.015 - x}$$

$$7.6\times10^{-3}\times(0.015 - x) = x^2$$

$$1.14\times10^{-4} - 7.6\times10^{-3}x - x^2 = 0$$

$$\frac{-b\pm\sqrt{b^2-4ac}}{2a} = \frac{7.6\times10^{-3}\pm\sqrt{\left(7.6\times10^{-3}\right)^2 + 4\times(1.14\times10^{-4})}}{-2}$$

$$\frac{7.6\times10^{-3}\pm(2.26\times10^{-2})}{-2}$$

$x = [H_3O^+] = 7.5\times10^{-3}$ M

pH = 2.12

(c) $H_2SO_3(aq) + H_2O(l) \rightleftharpoons HSO_3^-(aq) + H_3O^+(aq)$

$0.10 - x$	x	x

$$K_a = \frac{x^2}{0.10 - x} = 1.5\times10^{-2}$$

$$1.5\times10^{-2} = \frac{x^2}{0.10 - x}$$

$$1.5\times10^{-2}(0.10 - x) = x^2$$

$$1.5\times10^{-3} - 1.5\times10^{-2}x - x^2 = 0$$

$$\frac{-b \pm \sqrt{b^2 - 4ac}}{2a} = \frac{1.5\times10^{-2} \pm \sqrt{(1.5\times10^{-2})^2 + 4\times(1.5\times10^{-3})}}{-2}$$

$$\frac{1.5\times10^{-2} \pm 7.89\times10^{-2}}{-2}$$

$$x = [H_3O^+] = 3.2\times10^{-2}\,\text{M}$$

$$\text{pH} = 1.49$$

4.42 The first dissociation of tyrosine is

$$H_3Tyr^+(aq) + H_2O(l) \rightleftharpoons H_2Tyr(aq) + H_3O^+(aq)$$

$$K_{a1} = \frac{[H_2Tyr][H_3O^+]}{[H_3Tyr^+]}$$

$$\frac{[H_3Tyr^+]}{[H_2Tyr]} = \frac{[H_3O^+]}{6.31\times10^{-3}}$$

The relative concentration of tyrosine and its conjugate base at

(a) pH = 7 is 1.58×10^{-3}

(b) pH = 2.2 is 1.0

(c) pH = 1.5 is 5.01

4.43 (a) $(COOH)_2(aq) + H_2O(l) \rightleftharpoons HOOCCO_2^-(aq) + H_3O^+(aq)$

$0.15 - x$	x	x

$$K_a = \frac{x^2}{0.15 - x} = 5.9\times10^{-2}$$

$$5.9\times10^{-2} = \frac{x^2}{0.15 - x}$$

$$(5.9\times10^{-2})\times(0.15 - x) = x^2$$

$$8.85\times10^{-3} - (5.9\times10^{-2})x - x^2 = 0$$

$$\frac{-b \pm \sqrt{b^2 - 4ac}}{2a} = \frac{5.9\times10^{-2} \pm \sqrt{\left(5.9\times10^{-2}\right)^2 + 4\times(8.85\times10^{-3})}}{-2}$$

$$\frac{5.9\times10^{-2} \pm 0.197}{-2}$$

$$x = [H_3O^+] = [HOOCCO_2^-] = 6.91\times10^{-2}\ \text{M}$$

$$[(CO_2)_2^{2-}] = K_{a2} = 6.5\times10^{-5}\ \text{M}$$

$$[OH^-] = \frac{K_w}{6.91\times10^{-2}} = 1.45\times10^{-13}\ \text{M}$$

(b)

$$\text{pH} = \frac{1}{2}(\text{p}K_{a1} + \text{p}K_{a2})$$

$$\text{pH} = \frac{1}{2}(1.23 + 4.19)$$

$$\text{pH} = 2.71$$

4.44 $H_2S(aq) + H_2O(l) \rightleftharpoons HS^-(aq) + H_3O^+(aq)$

	H_2S	HS^-	H_3O^+
Initial concentration /(mol L^{-1})	0.15	0	0
Change to reach equilibrium /(mol L^{-1})	$-x$	$+x$	$+x$
Equilibrium concentration /(mol L^{-1})	$0.15 - x$	x	x

$$K_a = \frac{x^2}{0.065 - x} = 1.3\times10^{-7}$$

$$x = 9.19\times10^{-5}\ \text{M} = [H_3O^+] = [HS^-]$$

$$[H_2S] = 0.065\ \text{M}$$

$$[OH^-] = 1.1\times10^{-10}\ \text{M}$$

$$[S^{2-}] = K_{a2} = 7.1\times10^{-15}\ \text{M}$$

$K_{a2} = \dfrac{[S^{-2}][H_3O^+]}{[HS^-]}$, and since $[H_3O^+] = [HS^-]$

4.45 (a)

$H_2Gly^+(aq) + H_2O(l) \rightleftharpoons HGly(aq) + H_3O^+(aq)$

$HGly(aq) + H_2O(l) \rightleftharpoons Gly^-(aq) + H_3O^+(aq)$

	HGly	H_2Gly^+	Gly^-	H_3O^+
Initial []	Z	0	0	0
Change	$-(x+y)$	$+y$	$+x$	$+(x-y)$
$[\,]_{eq}$	$Z-x-y$	y	x	$x-y$

$$K_{a1} = \frac{[HGly][H_3O^+]}{[H_2Gly^+]} = \frac{(Z-x-y)(x-y)}{y}$$

$$K_{a2} = \frac{[Gly^-][H_3O^+]}{[HGly]} = \frac{x(x-y)}{Z-x-y}$$

$$K_{a1}y = Zx - Zy + x^2 + xy - yx + y^2 = Zx - Zy + x^2 + y^2$$

$$K_{a1}y \cong Z(x-y)$$

$$x \approx y$$

$$K_{a1}K_{a2} = \frac{x(x-y)^2}{y} \cong (x-y)^2 = [H_3O^+]^2$$

$$[H_3O^+] = (K_{a1}K_{a2})^{1/2}$$

$$pH = \frac{1}{2}(pK_{a1} + pK_{a2})$$

(b)

$H_3Asp^+(aq)+H_2O(l) \rightleftharpoons H_2Asp(aq)+H_3O^+(aq)$

$H_2Asp(aq)+H_2O(l) \rightleftharpoons HAsp^-(aq)+H_3O^+(aq)$

$HAsp^-(aq)+H_2O(l) \rightleftharpoons Asp^{2-}(aq)+H_3O^+(aq)$

	H_2Asp	H_3Asp^+	$HAsp^-$	H_3O^+
Initial []	Z	0	0	0
Change	$-(x+y)$	$+y$	$+x$	$+(x-y)$
$[\,]_{eq}$	$Z-x-y$	y	x	$x-y$

$$K_{a1} = \frac{[H_2Asp][H_3O^+]}{[H_3Asp^+]} = \frac{(Z-x-y)(x-y)}{y}$$

$$K_{a2} = \frac{[HAsp^-][H_3O^+]}{[H_2Asp]} = \frac{x(x-y)}{Z-x-y}$$

$$K_{a1}y = Zx - Zy + x^2 + xy - yx + y^2 = Zx - Zy + x^2 + y^2$$

$$K_{a1}y \cong Z(x-y)$$

$$x \approx y$$

$$K_{a1}K_{a2} = \frac{x(x-y)^2}{y} \cong (x-y)^2 = [H_3O^+]^2$$

$$[H_3O^+] = (K_{a1}K_{a2})^{1/2}$$

$$pH = \frac{1}{2}(pK_{a1} + pK_{a2})$$

(c)

$$H_3His^{2+}(aq) + H_2O(l) \rightleftharpoons H_2His^+(aq) + H_3O^+(aq)$$

$$H_2His^+(aq) + H_2O(l) \rightleftharpoons HHis(aq) + H_3O^+(aq)$$

$$HHis(aq) + H_2O(l) \rightleftharpoons His^-(aq) + H_3O^+(aq)$$

	HHis	H_2His^+	His^-	H_3O^+
Initial []	Z	0	0	0
Change	$-(x+y)$	$+y$	$+x$	$+(x-y)$
$[]_{eq}$	$Z-x-y$	y	x	$x-y$

$$K_{a2} = \frac{[HHis][H_3O^+]}{[H_2His^+]} = \frac{(Z-x-y)(x-y)}{y}$$

$$K_{a3} = \frac{[His^-][H_3O^+]}{[HHis]} = \frac{x(x-y)}{Z-x-y}$$

$$K_{a2}y = Zx - Zy + x^2 + xy - yx + y^2 = Zx - Zy + x^2 + y^2$$

$$K_{a2}y \cong Z(x-y)$$

$$x \approx y$$

$$K_{a2}K_{a3} = \frac{x(x-y)^2}{y} \cong (x-y)^2 = [H_3O^+]^2$$

$$[H_3O^+] = (K_{a2}K_{a3})^{1/2}$$

$$pH = \frac{1}{2}(pK_{a2} + pK_{a3})$$

4.46 The pH region for each buffer is (a) 3.08, (b) 4.19, (c) 12.7, (d) 6.59, (e) 6.03.

4.47 (a) H_3PO_4 and NaH_2PO_4
(b) NaH_2PO_4 and Na_2HPO_4

4.48 (a) at pH = p*K*a = 8.3

(b) $\text{TrisH}^+(aq) + OH^- \rightleftarrows \text{Tris}(aq) + H_2O(l)$

Assuming the concentration of Tris and its conjugate acid is 1 mol L^{-1}

	TrisH^+	Tris
Initial mol	0.1	0.1
After addition of 3.3 mmol NaOH	$0.1-3.3\times10^{-3}$	$0.1+3.3\times10^{-3}$
Concentration /(mol L^{-1})	0.97	1.03

$$\text{pH} = 8.30 + \log\frac{1.03}{0.97}$$

pH = 8.32

(c) $\text{Tris}(aq) + H_3O^+ \rightleftarrows \text{TrisH}^+(aq) + H_2O(l)$

Assuming the concentration of Tris and its conjugate acid is 1 mol L^{-1}

	Tris	TrisH^+
Initial mol in 100 mL	0.1	0.1
After addition of 6.0 mmol NaOH	$0.1-6.0\times10^{-3}$	$0.1+6.0\times10^{-3}$
Concentration /(mol L^{-1})	0.94	1.06

$$\text{pH} = 8.30 + \log\frac{0.94}{1.06}$$

pH = 8.25

Chapter 5: Thermodynamics of Ion and Electron Transport

5.8 The relation between ionic strength versus molalitiy of:

a) KCl

$$I = \frac{1}{2}\left((+1)^2 \times b + (-1)^2 \times b\right) = b$$

(b) $FeCl_3$

$$I = \frac{1}{2}\left((+3)^2 \times b + (-1)^2 \times 3 \times b\right) = 6b$$

(c) $CuSO_4$

$$I = \frac{1}{2}\left((+2)^2 \times b + (-2)^2 \times b\right) = 4b$$

5.9

$$I = \frac{1}{2}\left\{(+1)^2 \times (0.1) + (-1)^2 \times (0.1) + (+2)^2 \times (0.2) + (-2)^2 \times (0.2)\right\}$$

$$I = 0.9$$

5.10 (a) The ionic strength of the solution of $Ca(NO_3)_2$ and KNO_3 is

$$I = \frac{1}{2}\left((+1)^2 \times 0.150 + (-1)^2 \times 0.150 + (+2)^2 \times b + (-1)^2 \times 2b\right) = 0.250$$

$$0.500 = 0.300 + 6b$$

$$b = 0.333 \text{ mol kg}^{-1}$$

In 500 g of solvent, there are 0.166 mol of $Ca(NO_3)_2$. The mass of $Ca(NO_3)_2$ needed to prepare a $0.333 \text{ mol kg}^{-1}$ is then 27.2 g.

(b) The ionic strength of the solution of NaCl and KNO_3 is

$$I = \frac{1}{2}\left((+1)^2 \times 0.150 + (-1)^2 \times 0.150 + (+1)^2 \times b + (-1)^2 \times b\right) = 0.250$$

$$0.500 = 0.300 + 2b$$

$$b = 0.100 \text{ mol kg}^{-1}$$

In 500 g of solvent there are 0.05 mol of NaCl. The mass of NaCl needed to prepare a 0.100 mol kg^{-1} is then 2.92 g.

5.11 $\gamma_{\mp} = \left(\gamma_+ \gamma_-^2\right)^{1/3}$

5.12 We use Equation 5.4, $\log \gamma_{\pm} = -A|z_+ z_-| I^{1/2}$.

The ionic strength of the solution is:

$$I = \frac{1}{2}\left((+1)^2 \times 0.030 + (-1)^2 \times 0.030 + (+2)^2 \times 0.01 + (-1)^2 \times 2 \times 0.01\right)$$

$$I = 0.06$$

$$\log \gamma_{\pm} = -0.509|+2 \times -1| 0.06^{1/2}$$

$$\gamma_{\pm} = 0.56$$

5.13

$$\log \gamma_{\pm} = -\frac{0.509\sqrt{I}}{1 + B\sqrt{I}}$$

$$\frac{1}{\log \gamma_{\pm}} = -\frac{1 + B\sqrt{I}}{0.509\sqrt{I}} = -\frac{1}{0.509\sqrt{I}} - \frac{B}{0.509}$$

The intercept of a graph of $1/\log\gamma_{\pm}$ against $I^{1/2}$ is –3.972 giving a value of 2.02 for B.

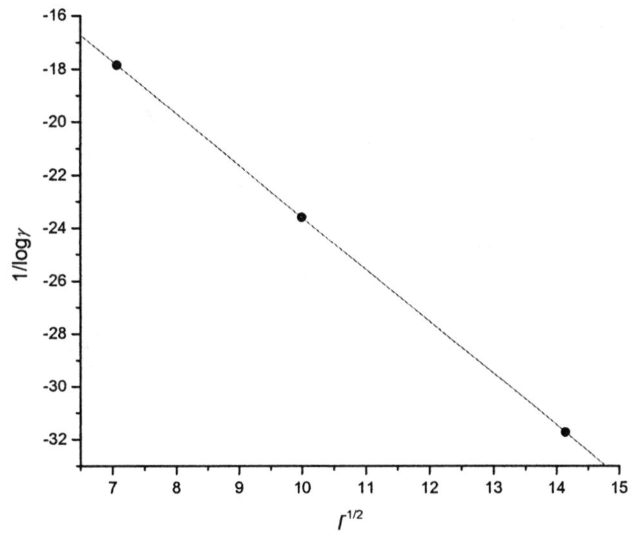

5.14 We calculate $\Delta_r G^{ATP}$ (see Equation 5.9).

$$\Delta_r G^{ATP} = \Delta_r G^{\ominus\prime} + RT \ln \frac{1}{100}$$

$$\Delta_r G^{ATP} = -31.3 \text{ kJ mol}^{-1} + \left(8.31447 \text{ J K}^{-1} \text{ mol}^{-1}\right) \times \left(310 \text{ K}\right) \times \ln \frac{1}{100}$$

$$\Delta_r G^{ATP} = -43.1 \text{ kJ mol}^{-1}$$

This energy is sufficient for the transport of Na^+ and K^+ ions.

5.15

$$\Delta\phi = \frac{RT}{F}\ln\left(\frac{\sum_i P_i[\mathrm{M}_i^+]_{\mathrm{out}} + \sum_j P_j[\mathrm{X}_j^-]_{\mathrm{in}}}{\sum_i P_i[\mathrm{M}_i^+]_{\mathrm{in}} + \sum_j P_j[\mathrm{X}_j^-]_{\mathrm{out}}}\right)$$

$$-3.0\times10^{-2}\ \mathrm{V} = \frac{(8.31447\ \mathrm{J\ K^{-1}mol^{-1}})\times(310\ \mathrm{K})}{9.648\times10^4\ \mathrm{C\ mol^{-1}}}\ln\left(\frac{(1.0\times5)+(P_{\mathrm{Na^+}}\times140)+(0.45\times10)}{(1.0\times100)+(P_{\mathrm{Na^+}}\times10)+(0.45\times100)}\right)$$

$$-1.1229 = \ln\left(\frac{(1.0\times5)+(P_{\mathrm{Na^+}}\times140)+(0.45\times10)}{(1.0\times100)+(P_{\mathrm{Na^+}}\times10)+(0.45\times100)}\right)$$

$$0.3253 = \frac{(1.0\times5)+(P_{\mathrm{Na^+}}\times140)+(0.45\times10)}{(1.0\times100)+(P_{\mathrm{Na^+}}\times10)+(0.45\times100)} = \frac{9.5+(P_{\mathrm{Na^+}}\times140)}{145+(P_{\mathrm{Na^+}}\times10)}$$

$$47.168-9.5 = (140-3.253)P_{\mathrm{Na^+}}$$

$$P_{\mathrm{Na^+}} = 0.27$$

5.16 Yes

5.17

$$\mathrm{CH_3COCOO^-(aq) + 2H^+(aq) + 2e^- \longrightarrow CH_3CHOHCOO^-(aq)}$$

$$\mathrm{NADH(aq) \longrightarrow NAD^+(aq) + H^+(aq) + 2e^-}$$

5.18

$$\mathrm{CH_3CH_2OH(aq) + NAD^+(aq) \longrightarrow CH_3CHO(aq) + NADH(aq) + H^+(aq)}$$

$$Q = \frac{[\mathrm{CH_3CHO}][\mathrm{NADH}][\mathrm{H^+}]}{[\mathrm{CH_3CH_2OH}][\mathrm{NAD^+}]}$$

$$\mathrm{NAD^+(aq) + H^+(aq) + 2e^- \longrightarrow NADH(aq)}$$

$$Q = \frac{[\mathrm{NADH}]}{[\mathrm{H^+}][\mathrm{NAD^+}]}$$

$$\mathrm{CH_3CHO(aq) + 2H^+(aq) + 2e^- \longrightarrow CH_3CH_2OH(aq)}$$

$$Q = \frac{[\mathrm{CH_3CH_2OH}]}{[\mathrm{CH_3CHO}][\mathrm{H^+}]^2}$$

5.19

$$2\,(HSCH_2CH(NH_2)COOH) \rightarrow$$
$$(HOOCCH(NH_2)CH_2SSCH_2CH(NH_2)COOH) + 2\,H^+ + 2e^-$$

5.20 The reduction half-reaction involving NADPH is:

$$NADP^+(aq) + H^+(aq) + 2e^- \longrightarrow NADPH(aq)$$

The overall reaction is:

$$2fd_{red}(aq) + NADP^+(aq) + 2\,H^+(aq) \longrightarrow 2fd_{ox}(aq) + NADPH(aq)$$

Subtraction reaction (1) from the overall equation is:

$$2fd_{red}(aq) + H^+(aq) - 2e^- \longrightarrow 2fd_{ox}(aq)$$

Or,

$$2fd_{ox}(aq) + 2e^- \longrightarrow 2fd_{red}(aq) + H^+(aq)$$

Two electrons are transferred in the reaction.

5.21 The two half-reactions are:

$$NAD^+(aq)+H^+(aq) + 2e^- \longrightarrow NADH(aq)$$

$$1/2\ O_2(g)+2H^+(aq) + 2e^- \longrightarrow H_2O(l)$$

Overall reaction: $NADH(aq) + 1/2\ O_2(g) + H^+(aq) \longrightarrow H_2O(l) + NAD^+(aq)$

$$E^{\oplus} = 1.14\ V$$

$$\Delta_r G^{\oplus} = -\nu FE^{\oplus} = -2\times(9.6485\times10^4\ C\ mol^{-1})\times(1.14\ V)$$

$$\Delta_r G^{\oplus} = -219\ kJ\ mol^{-1}$$

5.22

$$2\ cyt\ c_{red} + \frac{1}{2}O_2 + 2\,H^+ \rightarrow 2\ cyt\ c_{ox} + H_2O$$

$$\Delta_r G^{-} = -RT\ln K = -\nu FE^{-}$$

$$\Delta_r G^{-} = -\ 2\times(9.6485\times10^4\ C\ mol^{-1})\times(0.56\ V) = -108\ kJ\ mol^{-1}$$

$$\ln K = \frac{\nu FE^{-}}{RT} = \frac{2\times(9.6485\times10^4\ C\ mol^{-1})\times(0.56\ V)}{(8.31447\ J\ K^{-1}\ mol^{-1})\times(298\ K)} = 44$$

$$K = 1.28\times10^{19}$$

5.23

$$E = E^{-} - \frac{RT}{\nu F}\ln Q$$

$$E = 0 - \frac{(8.31447\ \text{J K}^{-1}\text{mol}^{-1})\times(298\ \text{K})}{2\times(9.6485\times10^{4}\ \text{C mol}^{-1})}\ln\frac{1.45}{\left(5\times10^{-3}\right)^{2}}$$

$$E = -0.1408\ \text{V}$$

$$E = 0 - \frac{(8.31447\ \text{J K}^{-1}\text{mol}^{-1})\times(298\ \text{K})}{2\times(9.6485\times10^{4}\ \text{C mol}^{-1})}\ln\frac{1.45}{\left(25\times10^{-3}\right)^{2}}$$

$$E = -0.0995\ \text{V}$$

The change in the electrode potential is 41mV.

5.24 The Nernst equation for the half-reaction is $2\,H^{+} + 2\,e^{-} \rightarrow H_2$.

$$E = E^{-} + \frac{RT}{2F}\ln a^{2}_{H^{+}}$$

$$E = -59.2\ \text{mV} \times \text{pH}$$

The pH of the $5.0\ \text{mmol L}^{-1}$ and $25.0\ \text{mmol L}^{-1}$ solutions are respectively 2.8 and 2.38.
The change in electrode potential, when the concentration changes from 5 to $25.0\ \text{mmol L}^{-1}$, is estimated to be 24.8 mV.

5.25 (a)

$$\text{Right}: 2H^{+}(aq) + 2e^{-} \longrightarrow H_2(g, p_R)$$

$$\text{Left}: 2H^{+}(aq) + 2e^{-} \longrightarrow H_2(g, p_L)$$

$$\text{R}-\text{L}:\ H_2(g, p_L) \longrightarrow H_2(g, p_R)$$

(b)

$$\text{Right: } Br_2(l) + 2e^{-} \longrightarrow 2Br^{-}(aq)$$

$$\text{Left: } Cl_2(g) + 2e^{-} \longrightarrow 2Cl^{-}(aq)$$

$$\text{R}-\text{L: } Br_2(l) + 2Cl^{-}(aq) \longrightarrow 2\,Br^{-}(aq) + Cl_2(g)$$

(c)

$$\text{Right: oxaloacetate(aq)} + 2H^{+}(aq) + 2e^{-} \longrightarrow \text{malate}^{2-}(aq)$$

$$\text{Left: NAD}^{+}(aq) + H^{+}(aq) + 2e^{-} \longrightarrow \text{NADH(aq)}$$

$$\text{R-L: oxaloacetate(aq)} + H^{+}(aq) + \text{NADH(aq)} \longrightarrow \text{NAD}^{+}(aq) + \text{malate}^{2-}(aq)$$

(d)

$$\text{Right: } MnO_2(s) + 4H^{+}(aq) + 2e^{-} \longrightarrow Mn^{2+}(aq) + 2H_2O(l)$$

$$\text{Left: } Fe^{2+}(aq) + 2e^{-} \longrightarrow Fe(s)$$

$$\text{R-L: } MnO_2(s) + 4H^{+}(aq) + Fe(s) \longrightarrow Mn^{2+}(aq) + 2H_2O(l) + Fe^{2+}(aq)$$

5.26 (a)

$$E = E^{-} - \frac{RT}{2F}\ln\frac{p_{H_2(R)}}{p_{H_2(L)}}$$

(b)

$$E = E^{-} - \frac{RT}{2F}\ln\frac{p_{Cl_2}a^2_{Br^-}}{p_{Br_2}a^2_{Cl^-}}$$

(c)

$$E = E^{-} - \frac{RT}{2F}\ln\frac{a_{NAD^+}a_{malate^{2-}}}{a_{NADH}a_{H^+}a_{oxaloacetate}}$$

(d)

$$E = E^{-} - \frac{RT}{2F}\ln\frac{a_{Mn^{2+}}a_{Fe^{2+}}}{a^4_{H^+}}$$

5.27 (a)

$Pt \mid CH_3CH_2OH(aq), CH_3CHO(aq), H^+(aq) \parallel NAD^+(aq), NADH(aq) \mid Pt$

$\nu = 2$

(b)

$Mg(s)|ATP^{4-}(aq), MgATP^{2-}(aq)||Mg^{2+}(aq)|Mg(s)$

$\nu = 2$

(c)

$Pt|Cyt\cdot c(red,aq), Cyt\cdot c(ox,aq) \parallel CH_3CH(OH)CO_2^-(aq), CH_3COCO_2^-(aq)|Pt$

$\nu = 2$

5.28 (a) $E^{\ominus} = -0.32\ V + 0.20\ V = -0.12\ V$

(b) $E^{\ominus} = -2.36\ V$

(c) $E^{\ominus} = -0.185\ V - 0.25\ V = -0.43\ V$

5.29 (a)

$MnO_2(s) + 4H^+(aq) + 2e^- \longrightarrow Mn^{2+}(aq) + 2H_2O(l)$

$$E = E^{\ominus} - \frac{RT}{\nu F}\ln Q$$

$$E = 1.51\ V - \frac{(8.31447\ J\ K^{-1}mol^{-1})\times(298\ K)}{5\times 9.6485\times 10^4\ C\ mol^{-1}}\ln\frac{1.0}{\left(1\times 10^{-6}\right)^8}$$

$$E = 0.94\ V$$

(b)

$$E = E^{-} - \frac{RT}{\nu F}\ln Q$$

$$E = 1.51\ \text{V} - \frac{(8.31447\ \text{J K}^{-1}\text{mol}^{-1})\times(298\ \text{K})}{5\times(9.6485\times10^{4}\ \text{C mol}^{-1})}\ln\frac{1.0}{\left[\text{H}_3\text{O}^{+}\right]^{8}}$$

$$E = 1.51 - (5.136\times10^{-3}\ \text{V})\times\ln\frac{1.0}{[\text{H}_3\text{O}^{+}]^{8}}$$

$$E = 1.51 + (0.0946\ \text{V})\times\left(-\log[\text{H}_3\text{O}^{+}]\right)$$

$$E = 1.51 - (0.0946\ \text{V})\times\text{pH}$$

5.30 In all cases the value of Q decreases, so the cell potential increases. In (c), H^+ is present in both half reactions (but there are 2 moles on the right side) so Q would indeed decrease (as indicated above in 5.26 c). Thus, cell potential would increase.

5.31 (a) $CH_3CH_2OH(aq) + NAD^+(aq) \rightarrow CH_3CHO(aq) + NADH(aq) + H^+(aq)$

$$E = E^{-} - \frac{RT}{2F}\ln\frac{a_{\text{CH}_3\text{CHO}}a_{\text{NADH}}a_{\text{H}^+}}{a_{\text{CH}_3\text{CH}_2\text{OH}}a_{\text{NAD}^+}}$$

The cell potential will increase when the pH of the solution is raised.

(b) $ATP^{4-}(aq) + Mg^{2+}(aq) \rightarrow MgATP^{2-}(aq)$

$$E = E^{-} - \frac{RT}{2F}\ln\frac{a_{\text{MgATP}^{2-}}}{a_{\text{ATP}^{4-}}a_{\text{Mg}^{2+}}}$$

The cell potential will increase when Mg^{2+} is added.

(c)

$$2\ \text{Cyt}\cdot\text{c(red, aq)} + \text{CH}_3\text{COCO}_2^-(\text{aq}) + 2\ \text{H}^+(\text{aq}) \rightarrow$$

$$2\ \text{Cyt}\cdot\text{c(ox, aq)} + \text{CH}_3\text{CH(OH)CO}_2^-(\text{aq})$$

$$E = E^{-} - \frac{RT}{2F}\ln\frac{a_{\text{Cyt}\cdot\text{c(ox)}}a_{\text{CH}_3\text{CH(OH)CO}_2^-}}{a_{\text{Cyt}\cdot\text{c(red)}}a_{\text{CH}_3\text{COCO}_2^-}a_{H^+}^{2}}$$

The cell potential will decrease when sodium lactate is added.

5.32 (a) The standard potential of the cell is -1.20 V

(b) $2Tl^+(aq) + Hg(s) \rightarrow Hg^{2+}(aq) + Tl(s)$

$$E = E^{-} - \frac{RT}{2F}\ln\frac{a_{\mathrm{H_g}^{2+}}}{a_{\mathrm{Tl}^{+}}^{2}}$$

$$E = -1.200\ \mathrm{V} - \frac{(8.31447\ \mathrm{J\ K^{-1}\ mol^{-1}})\times(298\ \mathrm{K})}{2\times(9.6485\times10^{4}\ \mathrm{C\ mol^{-1}})}\ln\frac{0.150}{0.93^{2}}$$

$$E = -1.178\ \mathrm{V}$$

5.33 (a) 2

$\mathrm{NADH(aq)} + \mathrm{O_2(g)} + 2\ \mathrm{H^+(aq)} \rightarrow 2\ \mathrm{NAD^+(aq)} + 2\ \mathrm{H_2O(l)} \quad \mathrm{E}^{\oplus} = +1.14\ \mathrm{V}$

(b)

$\mathrm{Malate^{2-}(aq)} + \mathrm{NAD^+(aq)} \rightarrow \mathrm{oxaloacetate^-(aq)} + \mathrm{NADH(aq)} + \mathrm{H^+(aq)} \quad \mathrm{E}^{\oplus} = -0.154$

(c) $\mathrm{O_2(g)} + 4\mathrm{H^+(aq)} + 4\ \mathrm{e^-} \rightarrow 2\ \mathrm{H_2O(l)} \quad \mathrm{E}^{\oplus} = +0.81\ \mathrm{V}$

(a)

$$\Delta_r G^{\oplus} = -\nu F E^{\oplus} = -2\times(9.6485\times10^{4}\ \mathrm{C\ mol^{-1}})\times(1.14\ \mathrm{V})$$

$$\Delta_r G^{\oplus} = -219\ \mathrm{kJ\ mol^{-1}}$$

(b)

$$\Delta_r G^{\oplus} = -\nu F E^{\oplus} = 2\times\left(9.6485\times10^{4}\mathrm{C\ mol^{-1}}\right)\times\left(0.154\ \mathrm{V}\right)$$

$$\Delta_r G^{\oplus} = +29.7\ \mathrm{kJ\ mol^{-1}}$$

(c)

$$\Delta_r G^{\oplus} = -\nu F E^{\oplus} = -4\times(9.6485\times10^{4}\ \mathrm{C\ mol^{-1}})\times(0.81\ \mathrm{V})$$

$$\Delta_r G^{\oplus} = -312\ \mathrm{kJ\ mol^{-1}}$$

5.34 (a) $\mathrm{AgCl(s)} + \mathrm{e^-} \rightarrow \mathrm{Ag(s)} + \mathrm{Cl^-(aq)}$

(b) The overall reaction is $\mathrm{Ag^+(aq)} + \mathrm{Cl^-(aq)} \rightarrow \mathrm{AgCl(s)}$

$$E = E^{-} - \frac{RT}{2F}\ln\frac{1}{a_{\mathrm{Cl^-}}a_{\mathrm{Ag^+}}}$$

$$E = 0.58\ \mathrm{V} - \frac{(8.31447\ \mathrm{J\ K^{-1}\ mol^{-1}})\times(298\ \mathrm{K})}{(9.6485\times10^{4}\ \mathrm{C\ mol^{-1}})}\ln\frac{1}{0.01\times0.025}$$

$$E = 0.37\ \mathrm{V}$$

5.35 (a)

$$E^{-} = 1.23\ \mathrm{V} + 0.34\ \mathrm{V} = 1.57\ \mathrm{V}$$

$$\Delta_r G^{-} = -\nu F E^{-} = -4\times(9.6485\times10^{4}\ \mathrm{C\ mol^{-1}})\times(1.57\ \mathrm{V})$$

$$\Delta_r G^{-} = -606\ \mathrm{kJ\ mol^{-1}}$$

$$\mathrm{O_2(g)} + 4\ \mathrm{H^+(aq)} + 4\ \mathrm{e^-} \rightarrow 2\ \mathrm{H_2O(l)}$$

$$2\ HSCH_2CH(NH_2)COOH \rightarrow$$
$$HOOCCH(NH_2)CH_2SSCH_2CH(NH2)COOH + 2\ H^+ + 2\ e^-)$$
$$4\ HSCH_2CH(NH_2)COOH + O_2(g) \rightarrow$$
$$2\ HOOCCH(NH_2)CH_2SSCH_2CH(NH_2)COOH) + 2\ H_2O(l)$$

$$\Delta_r H^{\ominus} = \left[\left(-2\times(285.83\ \text{kJ mol}^{-1})\right)+\left(-2\times(1032.7\ \text{kJ mol}^{-1})\right)\right]-\left[-4\times(534.1\ \text{kJ mol}^{-1})\right]$$

$$\Delta_r H^{\ominus} = -50.1\ \text{kJ mol}^{-1}$$

(b)

$$\Delta_r G^{\ominus} = -RT\ln K = -\nu FE^{\ominus}$$

$$\ln K = \frac{\nu FE^{\ominus}}{RT} = \frac{4\times(9.6485\times10^4\ \text{C mol}^{-1})\times(1.57\ \text{V})}{(8.31447\ \text{J K}^{-1}\text{mol}^{-1})\times(298\ \text{K})} = 244$$

$$\ln K' - \ln K = \frac{\Delta H}{R}\left(\frac{1}{T}-\frac{1}{T'}\right)$$

$$\ln K' = 244 - \frac{50.1\ \text{kJ mol}^{-1}}{\left(8.31447\times10^{-3}\ \text{kJ mol}^{-1}\text{K}^{-1}\right)}\left(\frac{1}{298\ \text{K}}-\frac{1}{308\ \text{K}}\right)$$

$$\ln K' = 243$$

$$\Delta_r G^{\ominus} = -RT\ln K' = -622\ \text{kJ mol}^{-1}$$

5.36

$$CH_3COCOO^- + 2H^+ + 2e^- \rightarrow CH_3CHOHCOO^-$$

$$E = E^{\ominus} - \frac{RT}{2F}\ln\frac{1}{a_{H^+}^2}$$

$$-0.19\text{V} = E - \frac{(8.31447\ \text{J K}^{-1}\ \text{mol}^{-1})\times(298\ \text{K})}{2\times(9.6485\times10^4\ \text{C mol}^{-1})}\ln\frac{1}{(1\times10^{-7})^2}$$

$$E = 0.22\ \text{V}$$

5.37

$$O_2 + 4\,H^+ + 4e^- \rightarrow 2\,H_2O$$

$$E^{\oplus} = E^{-} - \frac{RT}{\nu F}\ln Q$$

$$E^{\oplus} = 1.23 - \frac{(8.31447\ \text{J K}^{-1}\text{mol}^{-1})\times(298\ \text{K})}{4\times(9.6485\times10^4\ \text{C mol}^{-1})}\ln\frac{1.0}{\left(1\times10^{-7}\right)^4}$$

$$E^{\oplus} = 0.81\ \text{V}$$

$$\text{HOOCCH(NH}_2\text{)CH}_2\text{SSCH}_2\text{CH(NH}_2\text{)COOH)} + 2\,H^+ + 2e^- \rightarrow$$
$$2\ \text{HSCH}_2\text{CH(NH}_2\text{)COOH}$$

$$E^{\oplus} = E^{-} - \frac{RT}{\nu F}\ln Q$$

$$E^{\oplus} = -0.34 - \frac{(8.31447\ \text{J K}^{-1}\text{mol}^{-1})\times(298\ \text{K})}{2\times(9.6485\times10^4\ \text{C mol}^{-1})}\ln\frac{1.0}{\left(1\times10^{-7}\right)^2}$$

$$E^{\oplus} = -0.75\ \text{V}$$

Therefore, the emf of the cell is $E^{\oplus} = +1.56$ V.

5.38 (a) Yes
(b) No

5.39 Based on the standard reduction potentials:

Dehydroascorbic acid/Ascorbic acid	Glutathione, ox/ red	Lipoic acid, ox/red	Ubiquinone, ox/red
+0.08 V	−0.230 V	−0.29 V	+0.040 V

The most efficient is the one with the lower reduction potential: lipoic acid.

5.40

$$\Delta_r G^{-} = -RT\ln K = -\nu FE^{-}$$

$$\ln K = \frac{\nu FE^{-}}{RT} = -\frac{2\times(9.6485\times10^4\ \text{C mol}^{-1})\times(0.22\ \text{V})}{\left(8.31447\ \text{J K}^{-1}\ \text{mol}^{-1}\right)\times(298\ \text{K})} = -17.1$$

$$K = 3.6\times10^{-8}$$

5.41 (a)

$$\Delta_r G^{\ominus} = -RT \ln K = -\nu FE^{\ominus}$$

$$\ln K = \frac{\nu FE^{\ominus}}{RT} = \frac{2\times(9.6485\times10^{4}\ \mathrm{C\ mol^{-1}})\times E^{\ominus}}{(8.31447\ \mathrm{J\ K^{-1}mol^{-1}})\times(298\ \mathrm{K})} = 26.09$$

$$E^{\ominus} = +0.335\ \mathrm{V}$$

(b)

$$E^{\ominus} = 0.335\ \mathrm{V} = 0.254\ \mathrm{V} - E_{\mathrm{L}}^{\ominus}$$

$$E_{\mathrm{L}}^{\ominus} = +0.081\ \mathrm{V}$$

5.42 (a) The change in Gibbs energy for the reaction $HCO_3^- + H_2O + 2e^- \rightarrow CO_3^{2-} + H_2 + OH^-$ is calculated below:

$$\Delta_r G^{\ominus} = \left[\left(-527.81\ \mathrm{kJ\ mol^{-1}}\right)+\left(-157.24\ \mathrm{kJ\ mol^{-1}}\right)\right] - \left[\left(-586.77\ \mathrm{kJ\ mol^{-1}}\right) + \left(-237.13\ \mathrm{kJ\ mol^{-1}}\right)\right]$$

$$\Delta_r G^{\ominus} = 138.8\ \mathrm{kJ\ mol^{-1}}$$

Therefore, the standard potential of the HCO_3^-/CO_3^{2-}, H_2 couple is:

$$\Delta_r G^{\ominus} = -\nu FE^{0} = 2 \times 9.6485 \times 10^{4}\ \mathrm{C\ mol^{-1}} \times E^{\ominus} = 138.8 \times 10^{3}\ \mathrm{J\ mol^{-1}}$$

$$E^{\ominus} = +1.39\ \mathrm{V}$$

In (b) we have a simple hydrolysis of CO_3^-; no H_2, no oxidation number changing, and thus no electron transfer.

5.43

$$Cr_2O_7^{2-}(aq) + 14H^+(aq) + 6e^- \longrightarrow 2Cr^{3+}(aq) + 7H_2O(l)$$

$$E = E^{\ominus} - \frac{RT}{6F}\ln\frac{a_{Cr^{3+}}^2}{a_{Cr_2O_7^{2-}}\,a_{H^+}^{14}}$$

5.44 $AgCl(s) + \frac{1}{2} H_2(g) \rightarrow Ag(s) + Cl^-(aq) + H^+(aq)$

Assuming the a_{H^+} is the same as the a_{Cl^-} :

$$0.312 \text{ V} = 0.22 \text{ V} - \frac{(8.31447 \text{ J K}^{-1} \text{ mol}^{-1}) \times (298 \text{ K})}{9.6485 \times 10^4 \text{ C mol}^{-1}} \ln(a_{H^+})^2$$

$$0.312 \text{ V} = 0.22 \text{ V} - (0.0256 \text{ V}) \times \ln(a_{H^+})^2$$

$$0.092 = -(0.0256 \text{ V}) \times \ln(a_{H^+})^2$$

$$\ln(a_{H^+})^2 = -3.59$$

$$a_{H^+} = 0.166$$

$$\text{pH} = 0.78$$

5.45

$$\Delta G_m = F\Delta\phi - (RT \ln 10)\Delta\text{pH}$$

$$\Delta G_m = (9.6485 \times 10^4 \text{ C mol}^{-1}) \times (0.07 \text{ V}) + (8.31447 \text{ J mol}^{-1}\text{K}^{-1}) \times (298 \text{ K}) \times (\ln 10) \times 1.4$$

$$\Delta G_m = +14.74 \text{ kJ mol}^{-1}$$

The energy required for phosphorylation is 31 kJ mol^{-1}. The Gibbs energy available for phophorylation from the transport of 4 mol H^+ is:

$$\Delta G_m = (+14.74 \text{ kJ mol}^{-1}) \times 4 = +58.96 \text{ kJ mol}^{-1}$$

The number of mol ATP that could be synthesized is 58.96/31=1.90

5.46

$$\Delta G_m = F(\Delta\phi) - (RT \ln 10)(\Delta\text{pH})$$

$$\Delta G_m = (9.6485 \times 10^4 \text{ C mol}^{-1}) \times (0.07\text{V}) + (8.31447 \text{ J mol}^{-1}\text{K}^{-1}) \times (298 \text{ K}) \times (\ln 10) \times 1.3$$

$$\Delta G_m = +14.2 \text{ kJ mol}^{-1}$$

The energy required for phosphorylation is 31 kJ mol^{-1}. The Gibbs energy available for phophorylation from the transport of 2 moles of H^+ is:

$$\Delta G_m = +(14.2 \text{ kJ mol}^{-1}) \times 2 = +28.4 \text{ kJ mol}^{-1}$$

The number of moles ATP that could be synthesized is 28.4/31 = 0.91.

5.47 Yes. For example, fungus oxysporum reduces nitrate to ammonium and oxidizes ethanol to acetate to generate ATP.

5.48 (a)

$$E^{-} = 0.365\ \text{V} - 0.077\ \text{V}$$

$$E^{-} = 0.288\ \text{V}$$

$$\Delta_r G^{-} = -\nu F E^{-}$$

$$\Delta_r G^{-} = -(9.6485 \times 10^4\ \text{C mol}^{-1}) \times (0.288\ \text{V}) = -27.8\ \text{kJ mol}^{-1}$$

(b) A minimum of 8 electrons

Chapter 6: The Rates of Reactions

6.6

$$\frac{I_0}{I} = 10^A = \frac{100}{39.8} = 2.51$$

$$A = 0.399 = [\text{P450}]\varepsilon l$$

$$[\text{P450}] = \frac{0.399}{(291 \text{ L molcm}^{-1}) \times (0.65 \text{ cm})} = 2.1 \times 10^{-3} \text{ L mol}^{-1}$$

6.7

$$A_1 = \varepsilon_1[\text{A}]l + \varepsilon_1[\text{B}]l$$

$$A_2 = \varepsilon_{\text{A2}}[\text{A}]l + \varepsilon_{\text{B2}}[\text{B}]l$$

$$\frac{A_1}{\varepsilon_1 l} = [\text{A}] + [\text{B}]$$

$$[\text{A}] = \frac{A_1}{\varepsilon_1 l} - [\text{B}]$$

$$[\text{B}] = \frac{A_1}{\varepsilon_1 l} - [\text{A}]$$

Substitution of [A] into $A_2 = \varepsilon_{A2}[\text{A}]l + \varepsilon_{B2}[\text{B}]l$ gives:

$$A_2 = \varepsilon_{\text{A2}}[\text{A}]l + \varepsilon_{\text{B2}}[\text{B}]l$$

$$A_2 = \varepsilon_{\text{A2}} l \left(\frac{A_1}{\varepsilon_1 l} - [\text{B}] \right) + \varepsilon_{\text{B2}}[\text{B}]l$$

$$A_2 = \varepsilon_{\text{A2}} l \frac{A_1}{\varepsilon_1 l} - \varepsilon_{\text{A2}} l[\text{B}] + \varepsilon_{\text{B2}}[\text{B}]l$$

$$A_2 - \varepsilon_{\text{A2}} \frac{A_1}{\varepsilon_1} = \left(\varepsilon_{\text{B2}} l - \varepsilon_{\text{A2}} l\right)[\text{B}]$$

$$A_2\varepsilon_1 - \varepsilon_{\text{A2}} A_1 = \varepsilon_1 \left(\varepsilon_{\text{B2}} l - \varepsilon_{\text{A2}} l\right)[\text{B}]$$

$$[\text{B}] = \frac{A_2\varepsilon_1 - \varepsilon_{\text{A2}} A_1}{\left(\varepsilon_1\varepsilon_{\text{B2}} - \varepsilon_1\varepsilon_{\text{A2}}\right)l}$$

Substitution of [B] into $A_2 = \varepsilon_{A2}[A]l + \varepsilon_{B2}[B]l$ gives:

$$A_2 = \varepsilon_{A2}[A]l + \varepsilon_{B2}[B]l$$

$$A_2 = \varepsilon_{A2}[A]l + \varepsilon_{B2}l\left(\frac{A_1}{\varepsilon_1 l} - [A]\right)$$

$$A_2 = \varepsilon_{A2}[A]l + \varepsilon_{B2}\frac{A_1}{\varepsilon_1} - \varepsilon_{B2}l[A]$$

$$A_2 - \varepsilon_{B2}\frac{A_1}{\varepsilon_1} = (\varepsilon_{A2}l - \varepsilon_{B2}l)[A]$$

$$A_2\varepsilon_1 - \varepsilon_{B2}A_1 = \varepsilon_1(\varepsilon_{A2}l - \varepsilon_{B2}l)[A]$$

$$[A] = \frac{A_2\varepsilon_1 - \varepsilon_{B2}A_1}{(\varepsilon_1\varepsilon_{A2} - \varepsilon_1\varepsilon_{B2})l}$$

6.8

$$0.660 = (2.00\times10^3 \text{ L mol}^{-1}\text{ cm}^{-1})\times(1.00\text{ cm})\times[\text{Trp}] + (1.12\times10^4 \text{ L mol}^{-1}\text{ cm}^{-1})\times(1.00\text{ cm})\times[\text{Tyr}]$$

$$0.221 = (5.40\times10^3 \text{ L mol}^{-1}\text{ cm}^{-1})\times(1.00\text{ cm})\times[\text{Trp}] + (1.50\times10^3 \text{ L mol}^{-1}\text{ cm}^{-1})\times 1.00\text{ cm})\times[\text{Tyr}]$$

Solving for [Trp] in the first equation yields $3.3\times10^{-4} - 5.6\,[\text{Tyr}]$.
Substitution of [Trp] into the second equation yields:
$[\text{Tyr}] = 5.43\times10^{-5}\text{ mol L}^{-1}$, and $[\text{Trp}] = 2.58\times10^{-5}\text{ mol L}^{-1}$.

6.9

$$[A] = \frac{A_2\varepsilon_1 - \varepsilon_{B2}A_1}{(\varepsilon_1\varepsilon_{A2} - \varepsilon_1\varepsilon_{B2})l}$$

$$[\text{tryptophan}] = \frac{0.676\times(2.38\times10^3\text{ L mol}^{-1}\text{cm}^{-1}) - (1.58\times10^3\text{ L mol}^{-1}\text{cm}^{-1})\times0.468}{\begin{pmatrix}(2.38\times10^3\text{ L mol}^{-1}\text{cm}^{-1})\times(5.23\times10^3\text{ L mol}^{-1}\text{cm}^{-1}) - \\ (2.38\times10^3\text{ L mol}^{-1}\text{cm}^{-1})\times(1.58\times10^3\text{ L mol}^{-1}\text{cm}^{-1})\end{pmatrix}\times1.00\text{ cm}}$$

$$[\text{tryptophan}] = 1.00\times10^{-4}\text{ M}$$

$$0.468 = (2.28\times10^3\text{ L mol}^{-1}\text{cm}^{-1})\times(1.00\text{ cm})\times[\text{tryptophan}] + (2.28\times10^3\text{ Lmol}^{-1}\text{cm}^{-1})\times(1.00\text{ cm})[\text{tyrosine}]$$

$$0.468 = 0.228 + (2.28\times10^3)\times[\text{tyrosine}]$$

$$[\text{tyrosine}] = 9.65\times10^{-5}\text{ M}$$

6.10

$$v = \frac{1}{3}\frac{d[C]}{dt} = \frac{1}{2}\frac{d[D]}{dt} = -\frac{1}{2}\frac{d[A]}{dt} = -\frac{d[B]}{dt}$$

$$\frac{d[D]}{dt} = \frac{2}{3}\times 2.2 \text{ mol L}^{-1}\text{ s}^{-1} = 1.5 \text{ mol L}^{-1}\text{ s}^{-1}$$

$$-\frac{d[A]}{dt} = \frac{2}{3}\times 2.2 \text{ mol L}^{-1}\text{ s}^{-1} = 1.5 \text{ mol L}^{-1}\text{ s}^{-1}$$

$$-\frac{d[B]}{dt} = \frac{1}{3}\times 2.2 \text{ mol L}^{-1}\text{ s}^{-1} = 0.73 \text{ mol L}^{-1}\text{ s}^{-1}$$

6.11

$$r = k[A][B][C]$$

$$\text{M}\text{s}^{-1} = k\text{M}^{3}$$

$$k = \text{M}^{-2}\text{s}^{-1}$$

6.12 (a) For a second-order reaction:

$$v = k\ [A]^2 = \text{molecules m}^{-3}\text{s}^{-1} = k\times(\text{molecules m}^{-3})^2$$

$$k = \frac{\text{molecules m}^{-3}\text{ s}^{-1}}{(\text{molecules m}^{-3})^2} = \text{molecules}^{-1}\text{m}^3\text{s}^{-1}$$

For a third-order reaction:

$$k = \frac{\text{molecules m}^{-3}\text{ s}^{-1}}{(\text{molecules m}^{-3})^3} = \text{molecules}^{-2}\text{m}^6\text{s}^{-1}$$

(b) For a second-order reaction:

$$v = k\ [A]^2 = \text{kPa s}^{-1} = \text{k}\times(\text{kPa})^2$$

$$k = \frac{\text{kPa s}^{-1}}{(\text{kPa})^2} = \text{kPa}^{-1}\text{s}^{-1}$$

For a third-order reaction:

$$k = \frac{\text{kPa s}^{-1}}{(\text{kPa})^3} = \text{kPa}^{-2}\text{ s}^{-1}$$

6.13

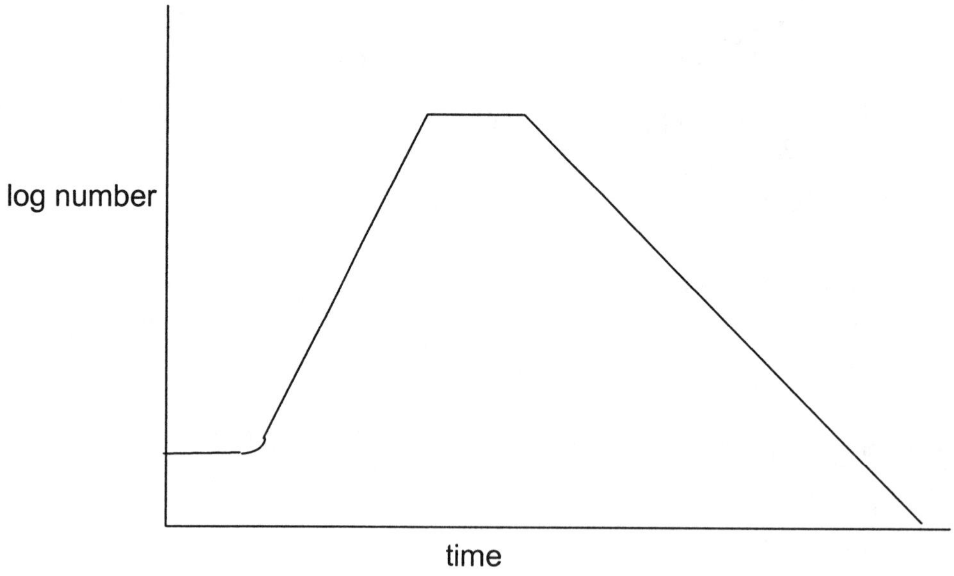

6.14

$k_{obs} = k\,[CO] = (5.8\times10^5\ \text{L mol}^{-1}\ \text{s}^{-1})\times(400\times10^{-3}\ \text{mol L}^{-1}) = 2.3\times10^5\ \text{s}^{-1}$

Using $[CO] = [CO]_0 e^{-kt}$ we could determine the change in CO concentration with time.

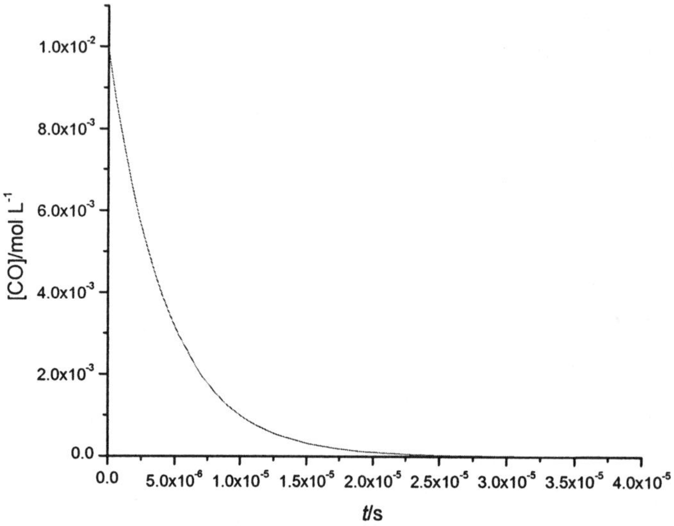

6.15

$$\frac{d[A]}{dt} = -k$$

$$\int_{[A]_0}^{[A]} d[A] = -k \int_0^t dt$$

$$[A] = [A]_0 - kt$$

$$\frac{[A]_0}{2} - [A]_0 = -kt$$

$$\frac{[A]_0}{2t} = k$$

$$k = 0.92 \text{ g L}^{-1}\text{h}^{-1}$$

6.16

$$\ln\frac{[A]}{[A]_0} = -\text{kt}$$

$$\ln\frac{56}{220} = -\text{k} \times 1.22 \times 10^4 \text{ s}$$

$$\text{k} = 1.12 \times 10^{-4} \text{ s}^{-1}$$

6.17

$$\frac{pV}{RT} = n_{CO_2} = \frac{(100 \text{ Pa})/(101.325\times10^3 \text{ Pa atm}^{-1})\times(0.250 \text{ L})}{(0.082 \text{ L atm K}^{-1}\text{mol}^{-1})\times(293 \text{ K})}$$

$$n_{CO_2} = 1.027\times10^{-5}$$

$$n_0(\text{pyruvate}) = 3.23\times10^{-4}$$

$$n_t(\text{pyruvate}) = 3.127\times10^{-4}$$

$$\ln\frac{3.127}{3.23} = -k\times522 \text{ s}$$

$$k = 6.19\times10^{-5}\text{s}^{-1}$$

6.18

$$\frac{1}{[A]} = \frac{1}{[A]_0} + kt$$

$$\frac{1}{0.0560 \text{ mol L}^{-1}} = \frac{1}{0.220 \text{ mol L}^{-1}} + k \times 1.22 \times 10^4 \text{ s}$$

$$13.3 \text{ mol}^{-1} \text{ L} = k \times 1.22 \times 10^4 \text{ s}$$

$$k = 1.09 \times 10^{-3} \text{ L mol}^{-1} \text{ s}^{-1}$$

6.19

$$\ln\frac{[A]}{[A]_0} = -kt$$

$$k = 1.12 \times 10^{-4} \text{s}^{-1}$$

6.20 The kinetics of NO + ½ Cl_2 → NOCl was followed under pseudo-second order conditions but the initial concentration of Cl_2 is nor given. In that case we are able to calculate the k_{obs}.

If we use p_{NO} = 300 Pa at t = 0, then after 522 s the pressure of NO is 200 Pa.

$$\frac{1}{[A]} = \frac{1}{[A]_0} + kt$$

$$\frac{1}{200 \text{ Pa}} = \frac{1}{300 \text{ Pa}} + k \times (522 \text{ s})$$

$$1.66 \times 10^{-3} \text{Pa}^{-1} = k \times 522 \text{ s}$$

$$k = 3.19 \times 10^{-6} \text{Pa}^{-1}\text{s}^{-1}$$

6.21 The rate law is $v = k[S]^n[E]^m$.

If enzyme concentrations are constant, the rate law takes the form:

$$v_0 = k'[S]_0^n$$

$$k' = k[E]_0^m$$

$$\log v_0 = \log k' + n \log[S]_0$$

The order of reaction with respect to S and the effective rate constant can be determined from the slope and the intercept of a graph of log v_0 against log$[S]_0$, respectively, as shown.

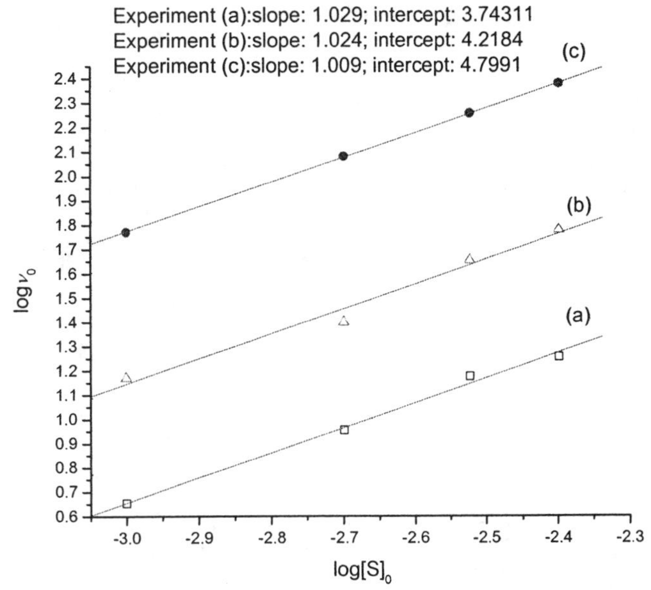

The slope of those graphs is 1, so the order with respect to S is first. The intercepts of the graphs can be used to determine the effective rate constant, k', for a given [E].

The order of reaction with respect to E and the rate constant can be determined from the slope and the intercept of a graph of log k' against $\log[E]_0$, respectively:

$$k' = k[E]_0^m$$

$$\log k' = \log k + m \log[E]_0$$

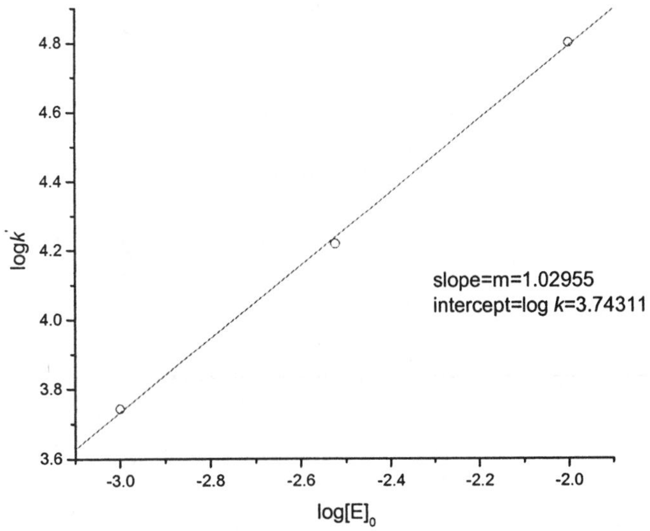

The slope of the graph is 1, so the order of reaction with respect to E is first. The intercept = log k is 3.74311, so the value of k is 5.535 M s^{-1}.

The rate law is then: $v = k[S][E]$.

6.22 A graph of ln[sucrose] versus time yields a straight line, indicating that the hydrolysis of sucrose is a first-order reaction.

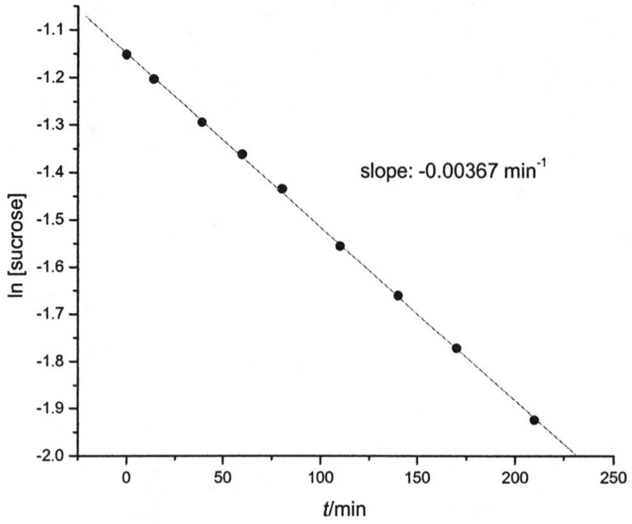

The slope of the graph is $-0.00367\ \text{min}^{-1}$, then $k = 0.00367\ \text{min}^{-1}$.

6.23 (a) The concentrations are exactly the same.
(b) The overall order of the reaction is second. See graph below:

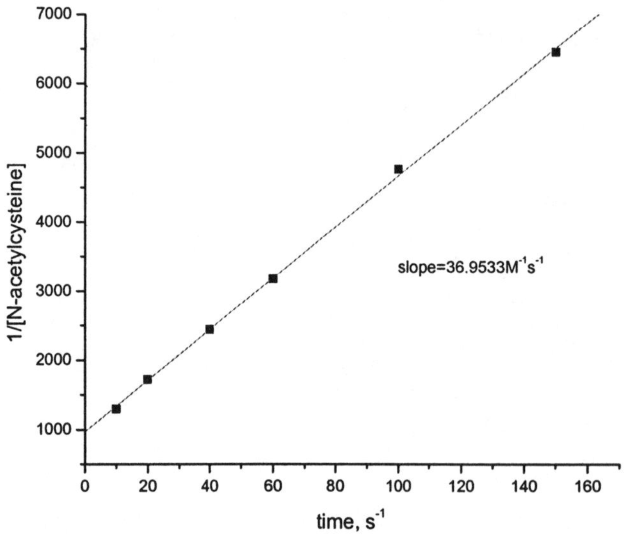

(c) The rate constant is $36.953\ \text{M}^{-1}\text{s}^{-1}$.

6.24 Results from Exercise 6.21a show that the reaction is a second-order reaction. If the order with respect to each of the reactants is first, then a graph of

$\ln\left(\dfrac{[B]/[B]_0}{[A]/[A]_0}\right)$ versus time should yield a straight line.

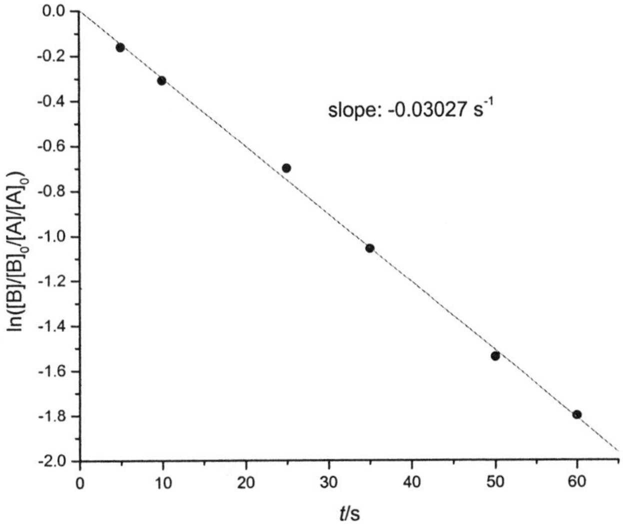

The slope is equal to $[B]_0 - [A]_0 \times k = -0.03027\ \text{s}^{-1}$.
We designated B to be N-acetylcysteine and A to be iodoacetamide. Since $[B]_0 - [A]_0$ is equal to $-1\times10^{-3}\text{mol L}^{-1}$, the value of $k = 30.27\ \text{L mol}^{-1}\ \text{s}^{-1}$.

6.25

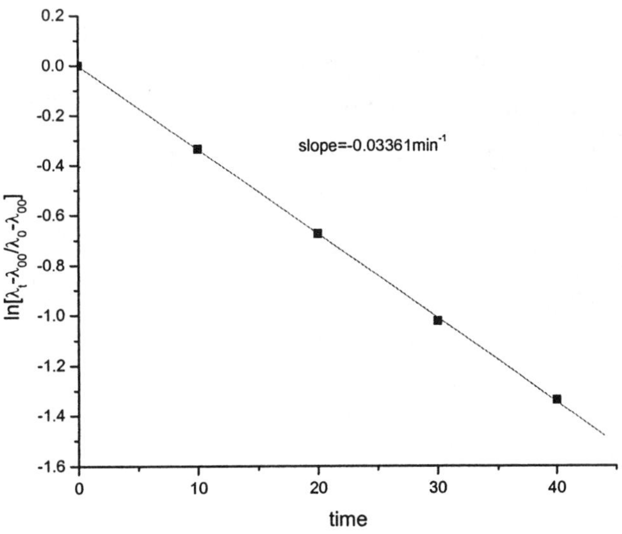

6.26 The integrated form of a third-order rate law is:

$$-\frac{d[A]}{dt} = k\,[A]^3$$

$$\int_{[A]_0}^{[A]} -\frac{d[A]}{[A]^3} = -k\int_0^t dt = -kt$$

$$-\frac{1}{2}\left(\frac{1}{[A]^2} - \frac{1}{[A]_0^{\,2}}\right) = -kt$$

$$\frac{1}{[A]^2} = \frac{1}{[A]_0^{\,2}} + 2kt$$

The plot should be $1/[A]^2$ versus time.

6.27

$$t_{1/2} = \frac{\ln 2}{k}$$

$$k = \frac{\ln 2}{t_{1/2}}$$

$$k = 3.14\times10^{-3}\,\mathrm{s}^{-1}$$

$$[A] = [A]_0 e^{-kt}$$

$$\frac{[A]_0}{64} = [A]_0\, e^{-kt}$$

$$\ln\frac{1}{64} = -kt$$

$$t = 1.32\times10^{3}\,\mathrm{s}$$

6.28 Using Equation 6.13, the rate constant is $1.209\times10^{-4}\ a^{-1}$. The age can be calculated using Equation 6.12c:

$$\ln(2) = kt \rightarrow k = \frac{\ln(2)}{5730a} = 1.209\times10^{-4}a^{-1}$$

$$\ln 0.69 = -\left(1.209\times10^{-4}\ a^{-1}\right)\times t$$

$$t = 3069\ a$$

6.29

$$t_{1/2} = \frac{\ln 2}{k}$$

$$k = \frac{\ln 2}{t_{1/2}}$$

$$k = 2.46 \times 10^{-2}\,\text{a}^{-1}$$

(a)

$$[\text{A}] = [\text{A}]_0 \text{e}^{-kt}$$

$$\frac{[\text{A}]}{[\text{A}]_0} = \text{e}^{-kt}$$

$$\frac{[\text{A}]}{1.00\ \mu\text{g}} = 0.62$$

(b)

$$\frac{[\text{A}]}{[\text{A}]_0} = \text{e}^{-kt}$$

$$\frac{[\text{A}]}{1.00\ \mu\text{g}} = 0.16$$

6.30 The rate constant for the reaction is: $(\ln 2)/\text{t}_{1/2} = 5.13 \times 10^{-6}\ a$.
We can calculate the time it takes for one P-O to break by using $[\text{A}] = [\text{A}]_0 \text{e}^{-kt}$. We find that it will take $1.95 \times 10^{-4}\ a$ or 1.50×10^{-9} half lives.

6.31

$$t_{1/2} = \frac{\ln 2}{k}$$

$$k = \frac{\ln 2}{t_{1/2}}$$

$$k = 0.15\ \text{h}^{-1}$$

$$\frac{[\text{A}]}{[\text{A}]_0} = \text{e}^{-kt}$$

$$\frac{[\text{A}]}{[\text{A}]_0} = 0.735$$

The mass of Phenobarbital that remains after 2 hours is $0.735 \times (30\ \text{mg/kg}) \times (15\ \text{kg}) = 328$ mg. To restore the initial amount $((30\ \text{mg/kg}) \times (15\ \text{kg}) = 450\text{mg})$, 120 mg must be re-injected.

6.32 To get an expression for $t_{1/2}$ in terms of n, we need to evaluate an integral such as:

$$\int_{[A]_0}^{[A]} \frac{d[A]}{[A]^n} = -k\int_0^t dt = -kt$$

$$\frac{1}{n-1}\left(\frac{1}{[A]^{n-1}} - \frac{1}{[A]_0^{\,n-1}}\right) = kt$$

An expression for $t_{1/2}$ is then:

$$\frac{1}{n-1}\left(\frac{2^{n-1}}{[A]_0^{\,n-1}} - \frac{1}{[A]_0^{\,n-1}}\right) = kt_{1/2}$$

$$\frac{1}{n-1}\left(\frac{2^{n-1} - 1}{[A]_0^{\,n-1}}\right) = kt_{1/2}$$

An expression for $t_{3/4}$ could be found by setting up $[A] = \frac{3}{4}[A]_0$

$$\frac{1}{n-1}\left(\frac{4^{n-1}}{3^{n-1}[A]_0^{\,n-1}} - \frac{1}{[A]_0^{\,n-1}}\right) = kt_{3/4}$$

$$\frac{1}{n-1}\left(\frac{\left(\frac{4}{3}\right)^{n-1} - 1}{[A]_0^{\,n-1}}\right) = kt_{3/4}$$

The ratio $t_{1/2}/t_{3/4}$ is then

$$t_{1/2}/t_{3/4} = \left(\frac{2^{n-1} - 1}{(4/3)^{n-1} - 1}\right)$$

6.33 (a)

$$\ln\left(\frac{[B]/[B]_0}{[A]/[A]_0}\right) = ([B]_0 - [A]_0)kt$$

$$\ln\frac{[B]}{[A]} = ([B]_0 - [A]_0)kt - \ln\frac{[B]_0}{[A]_0}$$

$$\ln\frac{[B]}{[A]} = (0.055 \text{ mol L}^{-1} - 0.150 \text{ mol L}^{-1}) \times (0.11 \text{ L mol}^{-1}\text{s}^{-1}) \times (15 \text{ s}) - \ln\frac{0.055}{0.150}$$

$$\ln\frac{[B]}{[A]} = -0.156 + 1.0033 = -1.159$$

$$\frac{[B]}{[A]} = 0.313 = \frac{[B]_0 - x}{[A]_0 - x}$$

$$x = 0.0117 \text{ mol L}^{-1}$$

$$[\text{ethylacetate}] = 0.150 \text{ mol L}^{-1} - 0.0117 \text{ mol L}^{-1} = 0.138 \text{ mol L}^{-1}$$

(b)

$$\ln\frac{[\text{B}]}{[\text{A}]} = (0.055\ \text{mol L}^{-1} - 0.150\ \text{mol L}^{-1}) \times (0.11\ \text{L mol}^{-1}\text{s}^{-1}) \times (900\ \text{s}) - \ln\frac{0.055}{0.150}$$

$$\ln\frac{[\text{B}]}{[\text{A}]} = -9.40 + 1.0033 = -8.40$$

$$\frac{[\text{B}]}{[\text{A}]} = 2.245 \times 10^{-4} = \frac{[\text{B}]_0 - x}{[\text{A}]_0 - x}$$

$$x = 0.0549\ \text{mol L}^{-1}$$

$$[\text{ethylacetate}] = 0.150\ \text{mol L}^{-1} - 0.0549\ \text{mol L}^{-1} = 0.095\ \text{mol L}^{-1}$$

6.34

$$\frac{1}{[\text{A}]} = \frac{1}{[\text{A}]_0} + kt$$

$$\frac{1}{0.026\ \text{mol L}^{-1}} = \frac{1}{0.260\ \text{mol L}^{-1}} + (1.24 \times 10^{-3}\text{L mol}^{-1}\ \text{s}^{-1}) \times t$$

$$34\ \ \text{mol}^{-1}\text{L} = t \times (1.24 \times 10^{-3}\text{L mol}^{-1}\ \text{s}^{-1})$$

$$t = 2.79 \times 10^{4}\text{s}$$

6.35

$$\ln\frac{k_1}{k_2} = \frac{E_\text{a}}{R}\left(\frac{1}{T_2} - \frac{1}{T_1}\right)$$

$$\ln\frac{1.78 \times 10^{-4}}{1.38 \times 10^{-3}} = \frac{E_\text{a}}{8.31447\ \text{J K}^{-1}\text{mol}^{-1}}\left(\frac{1}{310.15\ \text{K}} - \frac{1}{292.15\ \text{K}}\right)$$

$$2.048 = E_\text{a} \times (2.389 \times 10^{-5})$$

$$E_\text{a} = 85.7\ \text{kJ mol}^{-1}$$

$$k_1 = A\text{e}^{-E_\text{a}/RT}$$

$$A = 3.73 \times 10^{11}\ \text{L mol}^{-1}\text{s}^{-1}$$

6.36

$$\ln\frac{k_1}{k_2}=\frac{E_a}{R}\left(\frac{1}{T_2}-\frac{1}{T_1}\right)$$

$$\ln 1.1=\frac{408\times10^3\ \text{J mol}^{-1}}{8.31447\ \text{J K}^{-1}\ \text{mol}^{-1}}\left(\frac{1}{298\ \text{K}}-\frac{1}{T_1}\right)$$

$$0.0953=(4.91\times10^4)\times\left(\frac{1}{298\ \text{K}}-\frac{1}{T_1}\right)$$

$$T_1=298.2\ \text{K}$$

6.37 52 kJ mol^{-1}

6.38

$$\ln\frac{k_1}{k_2}=\frac{E_a}{R}\left(\frac{1}{T_2}-\frac{1}{T_1}\right)$$

$$\ln 1.23=\frac{E_a}{\left(8.31447\ \text{J K}^{-1}\ \text{mol}^{-1}\right)}\left(\frac{1}{293\ \text{K}}-\frac{1}{300\ \text{K}}\right)$$

$$0.207=E_a\times(9.57\times10^{-6})$$

$$E_a=21.6\ \text{kJ mol}^{-1}$$

6.39 The slope of a graph of ln k against $1/T$ (see below) will be used to determine the activation energy for the reaction.

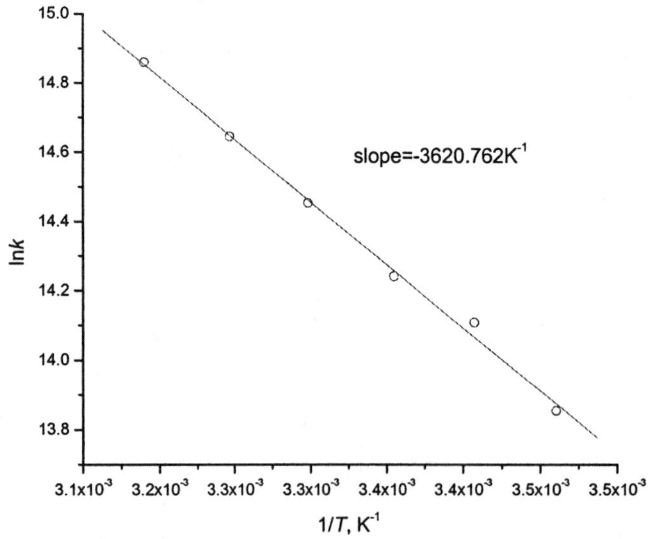

$$-3620.762\ \text{K}^{-1} = -\frac{\text{E}_\text{a}}{8.31447\ \text{J mol}^{-1}\text{K}^{-1}}$$

$$\text{E}_\text{a} = 30.1\ \text{kJ mol}^{-1}$$

6.40 Using Equation 6.22,

$$\ln\frac{k_1}{k_2} = \frac{E_\text{a}}{R}\left(\frac{1}{T_2} - \frac{1}{T_1}\right)$$

$$\ln\frac{40}{1} = \frac{E_\text{a}}{8.31447\ \text{J K}^{-1}\text{mol}^{-1}}\left(\frac{1}{277\ \text{K}} - \frac{1}{298\ \text{K}}\right)$$

$$3.68 = E_\text{a} \times (3.06 \times 10^{-5})$$

$$E_\text{a} = 120\ \text{kJ mol}^{-1}$$

6.41

$$\ln\frac{k_1}{k_2} = \frac{E_\text{a}}{R}\left(\frac{1}{T_2} - \frac{1}{T_1}\right)$$

$$\ln\frac{2}{1} = \frac{E_\text{a}}{\left(8.31447\ \text{J K}^{-1}\text{mol}^{-1}\right)}\left(\frac{1}{283.15\ \text{K}} - \frac{1}{293.15\ \text{K}}\right)$$

$$0.6931 = E_\text{a} \times \left(1.449 \times 10^{-5}\right)$$

$$E_\text{a} = 47.8\ \text{kJ mol}^{-1}$$

Chapter 7:
Accounting for the Rate Laws

7.8 In this case, the relation between the equilibrium constant and the rate constants is:

$$K = \frac{k_1}{k_2}$$

$$235 = \frac{7.4 \times 10^7 \text{ L mol}^{-1} \text{ s}^{-1}}{k_2}$$

$$k_2 = 3.1 \times 10^5 \text{ L mol}^{-1} \text{ s}^{-1}$$

7.9

$$A \underset{k_2}{\overset{k_1}{\rightleftarrows}} B$$

$$\frac{dx}{dt} = k_1([A]_0 - x) - k_2([B]_0 + x)$$

$$\frac{dx}{dt} = k_1[A]_0 - k_1 x - k_2[B]_0 - k_2 x$$

$$\frac{dx}{dt} = (k_1[A]_0 - k_2[B]_0) - (k_1 + k_2)x$$

$$\frac{dx}{dt} = (k_1 + k_2) \times \left[\frac{k_1[A]_0 - k_2[B]_0}{k_1 + k_2} - x\right]$$

$$m = \frac{k_1[A]_0 - k_2[B]_0}{k_1 + k_2}$$

$$\frac{dx}{dt} = (k_1 + k_2) \times (m - x)$$

$$\int_0^x \frac{dx}{m-x} = (k_1+k_2)\int_0^t dt$$

$$\ln \frac{k_1[A]_0 - k_2[B]_0}{k_1[A]_0 - k_2[B]_0 - x(k_1+k_2)} = (k_1+k_2)t$$

$$\frac{k_1[A]_0 - k_2[B]_0 - ([A]_0 - [A])(k_1+k_2)}{k_1[A]_0 - k_2[B]_0} = e^{-(k_1+k_2)t}$$

$$\frac{-k_2[B]_0 - [A]_0 k_2 + [A](k_1+k_2)}{k_1[A]_0 - k_2[B]_0} = e^{-(k_1+k_2)t}$$

$$[A] = \frac{e^{-(k_1+k_2)t}\left(k_1[A]_0 - k_2[B]_0\right) + [A]_0 k_2 + k_2[B]_0}{(k_1+k_2)}$$

if $[B]_0 = 0$

$$[A] = \frac{\left(e^{-(k_1+k_2)t} k_1 + k_2\right)[A]_0}{(k_1+k_2)}$$

7.10 (a) Following the steps described in Derivation 7.2:

$$H_2O(l) \underset{k_2}{\overset{k_1}{\rightleftarrows}} H^+(aq) + OH^-(aq)$$

$$\frac{d[H_2O]}{dt} = -k_1[H_2O] + k_2[H^+][OH^-]$$

$$\frac{d[H_2O]}{dt} = -k_1\left([H_2O]_{eq} + x\right) + k_2\left([H^+]_{eq} - x\right)\left([OH^-]_{eq} - x\right)$$

$$\frac{dx}{dt} = -\ k_1[H_2O]_{eq} - k_1x + k_2[H^+]_{eq}[OH^-]_{eq} + k_2x^2 - k_2[H^+]_{eq}x - k_2[OH^-]_{eq}x$$

at $\rightleftarrows$ $k_1[H_2O]_{eq} = k_2[H^+]_{eq}[OH^-]_{eq}$

$$\frac{dx}{dt} = -\ k_1x + k_2x^2 - k_2[H^+]_{eq}x - k_2[OH^-]_{eq}x$$

$$\frac{dx}{dt} = -\left(k_1 + k_2[H^+]_{eq} + k_2[OH^-]_{eq}\right)x$$

The term k_2x^2 has been neglected.

Integration of both sides yields:

$$\int_{x_0}^{x}\frac{dx}{x} = -(k_1 + k_2[H^+]_{eq} + k_2[OH^-]_{eq})\int_0^t dt$$

$$\ln\frac{x}{x_0} = -(k_1 + k_2[H^+]_{eq} + k_2[OH^-]_{eq})\,t$$

$$x = x_0 e^{-t/\tau}$$

$$\frac{1}{\tau} = k_1 + k_2\left([H^+]_{eq} + [OH^-]_{eq}\right)$$

(b) At equilibrium:

$$k_1[H_2O]_{eq} = k_2[H^+]_{eq}[OH^-]_{eq}$$

$$\frac{k_1}{k_2} = \frac{[H^+]_{eq}[OH^-]_{eq}}{[H_2O]_{eq}} = \frac{K_w}{55.5\ \text{mol L}^{-1}} = 1.76\times10^{-16}\ \text{mol L}^{-1}$$

Using the expression derived in part (a)

$$\frac{1}{\tau} = k_1 + k_2\left([H^+]_{eq} + [OH^-]_{eq}\right)$$

$$\frac{1}{37\times10^{-6}\ \text{s}} = k_2\times1.76\times10^{-16}\ \text{mol L}^{-1} + k_2\left(2\times\sqrt{9.77\times10^{-15}}\ \right)\text{mol L}^{-1}$$

$$k_2 = 1.37\times10^{11}\ \text{mol L}^{-1}\ \text{s}^{-1}$$

The rate constant for the forward direction is then easily obtained:

$$k_1 = 2.4\times10^{-5}\ \text{s}^{-1}$$

7.11

$$2A \underset{k_b}{\overset{k_a}{\rightleftarrows}} A_2$$

$$\frac{d[A]}{dt} = -k_a[A]^2 + k_b[A_2]$$

$$\frac{dx}{dt} = -k_a\left(2x + [A]_{eq}\right)^2 + k_b\left([A_2]_{eq} - x\right)$$

$$\frac{dx}{dt} = -4x^2k_a - 4x[A]_{eq}k_a - k_a[A]^2_{eq} + k_b[A_2]_{eq} - k_bx$$

At equilibrium: $k_a[A]^2_{eq} = k_b[A_2]_{eq}$

$$\frac{dx}{dt} = -4x^2k_a - 4x[A]_{eq}k_a - k_bx$$

The term $4x^2k_a$ will be neglected.

$$\frac{dx}{dt} = -4x[A]_{eq}k_a - k_b x$$

$$\div x$$

$$\int_{x_0}^{x}\frac{dx}{x} = -(4[A]_{eq}k_a + k_b)\int_0^t dt$$

$$\ln\frac{x}{x_0} = -(4[A]_{eq}k_a + k_b)t$$

$$\tau = \frac{1}{4[A]_{eq}k_a + k_b}$$

7.12 (a)

$$\frac{1}{\tau} = 4[A]_{eq}k_a + k_b$$

$$\frac{1}{\tau^2} = \left(4[A]_{eq}k_a + k_b\right)^2$$

$$\frac{1}{\tau^2} = 16[A]^2{}_{eq}k_a^2 + k_b{}^2 + 8[A]_{eq}k_a k_b$$

$$\frac{1}{\tau^2} = 16$$

(b) The construction of a graph of $1/\tau^2$ versus $[A]_{tot}$ will allow us to evaluate the product $k_a k_b$ from the slope and k_b from the intercept.

7.13

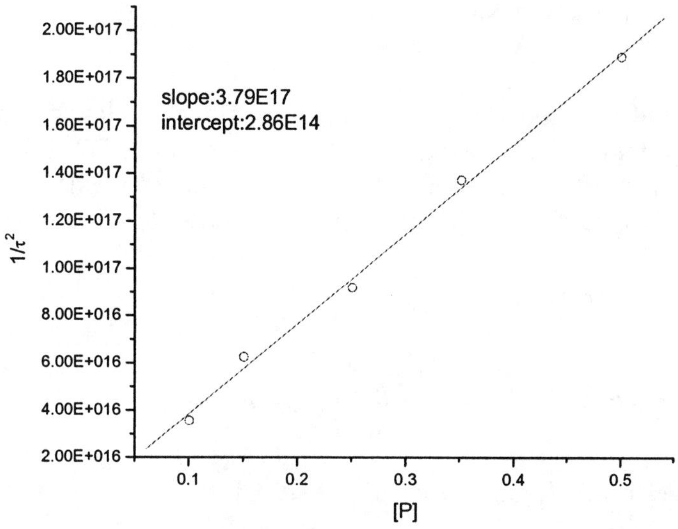

The value of the intercept can be used to determine $k_b = 1.69 \times 10^7$ s^{-1}. The slope and the value of k_b are used to obtain the value of $k_a = 2.80 \times 10^9$ s^{-1}.

$$K = \frac{k_a}{k_b} = 166$$

7.14 As described in Section 7.4, the rate of formation of P is:

(i) $\frac{d[P]}{dt} = k_2[I]$

The net rate of formation of I is:

(ii) $\frac{d[I]}{dt} = k_1[A] - k_2[I]$

And the rate at which A decays is:

(iii) $\frac{d[A]}{dt} = -k_1[A]$

Differentiation of Expression 7.7 yields:

$$\frac{d[A]}{dt} = -[A]_0 k_1 e^{-k_1 t} = -k_1[A]$$

which is the 1st order rate law (iii).

Differentiation of Expression 7.9 yields:

$$\text{(iv)}\quad \frac{d[I]}{dt} = \frac{k_1}{k_2 - k_1}\left(-k_1 e^{-k_1 t} + k_2 e^{-k_2 t}\right)[A]_0$$

In this case, to confirm Expression 7.9 is a solution of the rate law given by (ii) is accomplished by doing the following: (a) substitute Equations 7.7 and 7.9 into (ii), and (b) obtain a differential equation identical to the one yielded by differentiation of Expression 7.9 (Expression (iv)). These steps are shown below:

$$\frac{d[I]}{dt} = k_1[A]_0 e^{-k_1 t} - \frac{k_1 k_2}{k_2 - k_1}\left(e^{-k_1 t} - e^{-k_2 t}\right)[A]_0$$

$$\frac{d[I]}{dt} = \frac{k_1}{k_2 - k_1}\left(k_2 - k_1\right)[A]_0 e^{-k_1 t} - \frac{k_1 k_2}{k_2 - k_1}\left(e^{-k_1 t} - e^{-k_2 t}\right)[A]_0$$

$$\frac{d[I]}{dt} = \frac{k_1}{k_2 - k_1}[A]_0\left[\left(k_2 - k_1\right)e^{-k_1 t} - k_2\left(e^{-k_1 t} - e^{-k_2 t}\right)\right]$$

$$\frac{d[I]}{dt} = \frac{k_1}{k_2 - k_1}[A]_0\left[\left(k_2 e^{-k_1 t} - k_1 e^{-k_1 t}\right) - \left(k_2 e^{-k_1 t} - k_2 e^{-k_2 t}\right)\right]$$

$$\frac{d[I]}{dt} = \frac{k_1}{k_2 - k_1}[A]_0\left[k_2 e^{-k_1 t} - k_1 e^{-k_1 t} - k_2 e^{-k_1 t} + k_2 e^{-k_2 t}\right]$$

$$\frac{d[I]}{dt} = \frac{k_1}{k_2 - k_1}[A]_0\left[-k_1 e^{-k_1 t} + k_2 e^{-k_2 t}\right]$$

This is identical to expression (iv).

Differentiation of Expression 7.10 yields:

$$\frac{d[P]}{dt} = \frac{1}{k_2 - k_1}\left(-k_1 k_2 e^{-k_2 t} + k_1 k_2 e^{-k_1 t}\right)[A]_0$$

$$\frac{d[P]}{dt} = \frac{k_1 k_2}{k_2 - k_1}\left(e^{-k_1 t} - e^{-k_2 t}\right)[A]_0$$

$$\frac{d[P]}{dt} = k_2[I]$$

This is the 1st order rate law (i).

7.15

$$t_{1/2} = \frac{\ln 2}{k}$$

$$k_1 = \frac{\ln 2}{22.5 \text{ d}} = 3.08 \times 10^{-2} \text{ d}^{-1}$$

$$k_2 = \frac{\ln 2}{33.0 \text{ d}} = 2.10 \times 10^{-2} \text{ d}^{-1}$$

$$t = \frac{1}{k_1 - k_2} \ln \frac{k_1}{k_2}$$

$$t = \frac{1}{(3.08 \times 10^{-2} \text{d}^{-1}) - (2.10 \times 10^{-2} \text{d}^{-1})} \ln \frac{3.08 \times 10^{-2}}{2.10 \times 10^{-2}}$$

$$t = 39.1 \text{ d}$$

7.16 The graph below shows the [I] versus time for k_2/k_1 equal to 0.1, 2, 5, and 25.

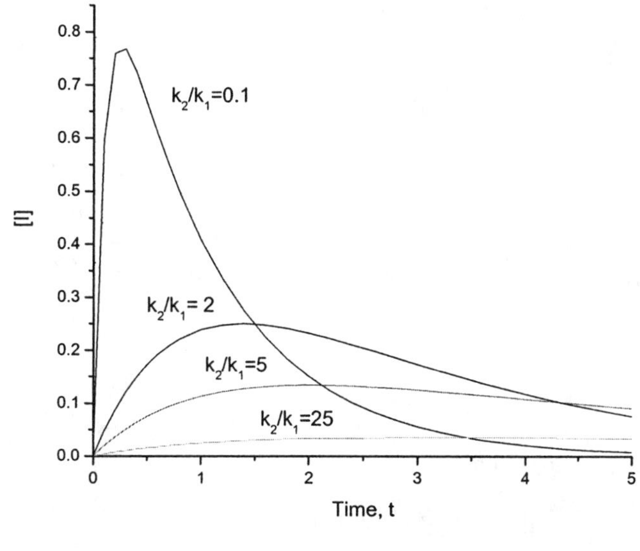

As k_2 increases or k_1 decreases, the concentration of the intermediate does not change significantly with time. Hence, the steady state approximation becomes increasingly valid with increasing k_2.

7.17 The first step of the mechanism is rate-determining, so the rate of the overall reaction is equal to the rate of the first step. The order of the reaction is first with respect to H_2O_2 and first with respect to Br^-.

7.18 The equilibrium is very fast, so:

$$\frac{d[P]}{dt} = k_2[A][B]$$

$$K = \frac{[A]^2}{[A_2]} = \frac{k_1}{k_{-1}}$$

$$[A] = \sqrt{\frac{k_1}{k_{-1}}}[A_2]^{1/2}$$

$$\frac{d[P]}{dt} = k_2\left(\frac{k_1}{k_{-1}}\right)^{1/2}[A_2]^{1/2}[B]$$

7.19 Unstable helix will be identified with I. The rate constants for the first step are k_1 and k_{-1} for the forward and reverse reactions, respectively. The rate constant for the slow reaction is k_2.

$$\frac{d[P]}{dt} = k_2[I]$$

$$K = \frac{[I]}{[A][B]} = \frac{k_1}{k_{-1}}$$

$$[I] = \frac{k_1}{k_{-1}}[A][B]$$

$$\frac{d[P]}{dt} = k_2\frac{k_1}{k_{-1}}[A][B]$$

7.20 Using the steady-state approximation,

$$\frac{d[O_3]}{dt} = -\ k_1[O_3] + k_{-1}[O_2][O] - k_2[O][O_3]$$

$$\frac{d[O]}{dt} = k_1[O_3] - k_{-1}[O_2][O] - k_2[O][O_3] = 0$$

$$[O] = \frac{k_1[O_3]}{k_{-1}[O_2] + k_2[O_3]}$$

Then sbstitute [O] into the expression of $d[O_3]/dt$:

$$\frac{d[O_3]}{dt} = -k_1[O_3] + \frac{k_{-1}k_1[O_2][O_3]}{k_{-1}[O_2]+k_2[O_3]} - \frac{k_2k_1[O_3]^2}{k_{-1}[O_2]+k_2[O_3]}$$

$$\frac{d[O_3]}{dt} = \frac{-k_1[O_3](k_{-1}[O_2]+k_2[O_3]) + k_{-1}k_1[O_2][O_3] - k_2k_1[O_3]^2}{k_{-1}[O_2]+k_2[O_3]}$$

$$\frac{d[O_3]}{dt} = \frac{-k_{-1}k_1[O_3][O_2] - k_1k_2[O_3]^2 + k_{-1}k_1[O_2][O_3] - k_2k_1[O_3]^2}{k_{-1}[O_2]+k_2[O_3]}$$

$$\frac{d[O_3]}{dt} = \frac{-2\,k_1k_2[O_3]^2}{k_{-1}[O_2]+k_2[O_3]}$$

If step 2 is slow, the $d[O_3]/dt$ reduces to:

$$\frac{d[O_3]}{dt} = -\frac{2\,k_1k_2[O_3]^2}{k_{-1}[O_2]}$$

7.21

$$\frac{d[P]}{dt} = k_3[AH][A^-]$$

$$\frac{d[A^-]}{dt} = k_1[AH][B] - k_2[A^-][BH^+] - k_3[A^-][AH] = 0$$

$$k_1[AH][B] = k_2[A^-][BH^+] + k_3[A^-][AH]$$

$$[A^-] = \frac{k_1[AH][B]}{k_2[BH^+]+k_3[AH]}$$

$$\frac{d[P]}{dt} = \frac{k_1k_3[AH]^2[B]}{k_2[BH^+]+k_3[AH]}$$

7.22 Considering the equilibrium is very fast,

$$\frac{d[BH^+]}{dt} = k_2[HAH^+][B]$$

$$K = \frac{[HAH^+]}{[HA][H^+]} = \frac{k_1}{k_{-1}}$$

$$[HAH^+] = \frac{k_1}{k_{-1}}[HA][H^+]$$

$$\frac{d[BH^+]}{dt} = k_2\frac{k_1}{k_{-1}}[HA][H^+][B]$$

If the source for the H^+ is HA, then:

$$K_a = \frac{[H^+][A^-]}{[HA]} = \frac{[H^+]^2}{[HA]}$$

Solve for $[H^+]$ and Substitute into the expression for $d[BH^+]/dt$. This yields:

$$\frac{d[BH^+]}{dt} = K_a^{1/2}\, k_2 \frac{k_1}{k_{-1}} [HA]^{3/2}[B]$$

7.23

$$\frac{dN}{dt} = nB - dD$$

$$\int_{N_0}^{N} \frac{dN}{bN - dN} = \int_0^t dt$$

$$\int_{N_0}^{N} \frac{dN}{N} = (b-d)\int_0^t dt$$

$$\ln\frac{N}{N_0} = (b-d)t$$

$$\frac{N}{N_0} = e^{(b-d)t}$$

A graph of ln *N* against time is shown below

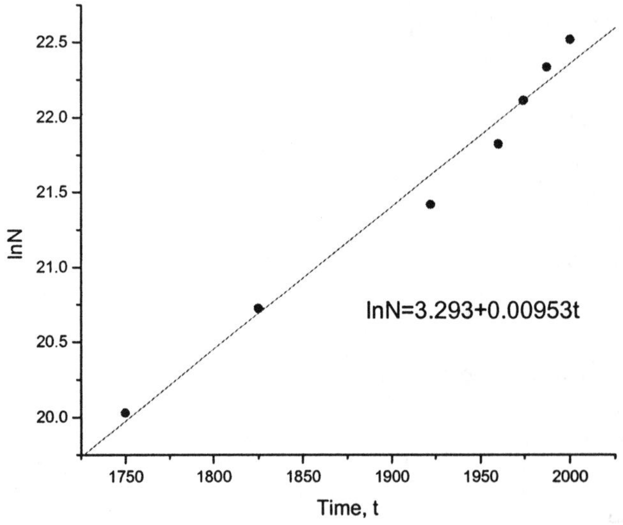

This model as a whole fits the data quite well.

7.24 Using Equation 7.22 and the relation 1 J = kg m^2 s^{-2}

$$k_d = \frac{8RT}{3\eta}$$

$$k_d = \frac{8 \times 8.31477 \text{ J K}^{-1} \text{ mol}^{-1} \times 298 \text{ K}}{3 \times 1.06 \times 10^{-3} \text{ kg m}^{-1} \text{ s}^{-1}}$$

$$k_d = 6.23 \times 10^6 \text{ m}^3 \text{ mol}^{-1} \text{ s}^{-1}$$

7.25 The ratio will increase because the reaction with the greater activation energy will be more sensitive to temperature.

7.26 We can find the ratio of rates of catalyzed to non-catalyzed reactions by using Equation 7.29.

$$\frac{k_{cat}}{k_{uncat}} = \exp(-\Delta^{\ddagger}G/RT)_{cat} / \exp(-\Delta^{\ddagger}G/RT)_{uncat} = \exp((-\Delta^{\ddagger}G_{cat} + \Delta^{\ddagger}G_{uncat})/RT)$$

$$\frac{k_{cat}}{k_{uncat}} = e^{\frac{90 \times 10^3 \text{ J mol}^{-1}}{8.31447 \text{ J K}^{-1} \text{ mol}^{-1} \times 310 \text{ K}}}$$

$$\frac{k_{cat}}{k_{uncat}} = 1.46 \times 10^{15}$$

7.27

$$A = \sigma \left(\frac{8kT}{\pi\mu}\right)^{1/2} N_A$$

$$\mu = \frac{m_{H_2} m_{C_2H_4}}{m_{H_2} + m_{C_2H_4}} = \left(\frac{2.016 \text{ u} \times 28.05 \text{ u}}{2.016 \text{ u} + 28.05 \text{ u}}\right) \times (1.66 \times 10^{-27} \text{ kg u}^{-1}) = 3.12 \times 10^{-27} \text{kg}$$

$$A = \frac{1}{2} \times (0.64 + 0.27) \text{nm}^2 \times \left(\frac{10^{-9} \text{m}}{1 \text{ nm}}\right)^2 \left(\frac{8 \times (1.380 \times 10^{-23} \text{ J K}^{-1}) \times (673.15 \text{ K})}{3.14 \times (3.12 \times 10^{-27} \text{kg})}\right)^{1/2} \times 6.02 \times 10^{23} \text{mol}^{-1}$$

$$A = \left(7.54 \times 10^8 \text{m}^3 \text{s}^{-1} \text{mol}^{-1}\right) \times \left(\frac{1000 \text{ dm}^3}{1 \text{ m}^3}\right)$$

$$A = 7.54 \times 10^{11} \text{L mol}^{-1} \text{s}^{-1}$$

7.28 It seems that in the case of the frog, the transformation from metarhodopsin I to matarhodospin II has lower activation energy than in the bovine case. If the rate of transformation from metarhodopsin I to matarhodospin II does not change dramatically with temperature, frogs could enhance their chance of survival in both cold and warmer environment.

7.29 $$\ln\frac{k_1}{k_2} = \Delta^{\ddagger}G/R\left(\frac{1}{T_2}-\frac{1}{T_1}\right)$$

$$\ln\frac{1.2\times10^{-7}}{4.6\times10^{-7}} = \Delta^{\ddagger}G/8.31447\text{ J K}^{-1}\text{mol}^{-1}\left(\frac{1}{343.15\text{ K}}-\frac{1}{333.15\text{ K}}\right)$$

$$1.344 = \Delta^{\ddagger}G\times(1.052\times10^{-5})$$

$$\Delta^{\ddagger}G = 127\text{ kJ mol}^{-1}$$

7.30 A graph of ln(k/T) against 1/T yields an intercept equal to 23.3 K.

$$23.3 = \ln\frac{\text{k}}{\text{h}} + \Delta^{\ddagger}S/\text{R}$$

$$\Delta^{\ddagger}S = \text{R}\left(23.3 - \ln\frac{\text{k}}{\text{h}}\right)$$

$$\Delta^{\ddagger}S = 8.31447\text{ J mol}^{-1}\text{ K}^{-1}\left(23.3 - \ln\frac{1.381\times10^{-23}}{6.626\times10^{-34}}\right)$$

$$\Delta^{\ddagger}S = -3.83\text{ J mol}^{-1}\text{ K}^{-1}$$

7.31

$$\ln k_{TS} = \ln\frac{kT}{h} - \Delta^{\ddagger}G/RT$$

$$\ln\frac{k_{TS}}{T} = \ln\frac{k}{h} - (\Delta^{\ddagger}H - T\Delta^{\ddagger}S)/RT$$

$$\ln\frac{k_{TS}}{T} = \ln\frac{k}{h} + \Delta^{\ddagger}S/R - \Delta^{\ddagger}H/RT$$

A graph of ln(*k*/*T*) against 1/*T* is shown below:

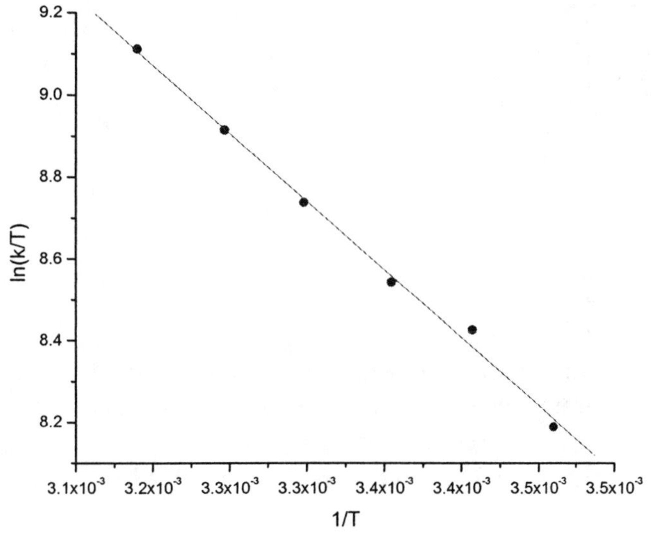

$$\text{Slope} = -\Delta^{\ddagger}H/R = -3319.89 \text{ K}$$

$$\Delta^{\ddagger}H = (3319.89 \text{ K}) \times (8.31447 \text{ J mol}^{-1} \text{ K}^{-1}) = 27.6 \text{ kJ mol}^{-1}$$

$$\text{intercept} = \ln\frac{k}{h} + \Delta^{\ddagger}S/R = 19.69$$

$$\Delta^{\ddagger}S = \text{R}\left(19.69 - \ln\frac{k}{h}\right)$$

$$\Delta^{\ddagger}S = \left(8.31447 \text{ J mol}^{-1}\text{K}^{-1}\right)\left(19.69 - \ln\frac{1.381\times10^{-23}}{6.626\times10^{-34}}\right)$$

$$\Delta^{\ddagger}S = -33.8 \text{ J mol}^{-1}\text{K}^{-1}$$

$$\Delta^{\ddagger}G = \left(27.6 \text{ kJ mol}^{-1}\right) + \left(33.8\times10^{-3} \text{ kJ mol}^{-1}\text{K}^{-1}\right)\times\left(300\ K\right)$$

$$\Delta^{\ddagger}G = 37.74 \text{ kJ mol}^{-1}$$

7.32 (a) A reaction profile for this reaction is:

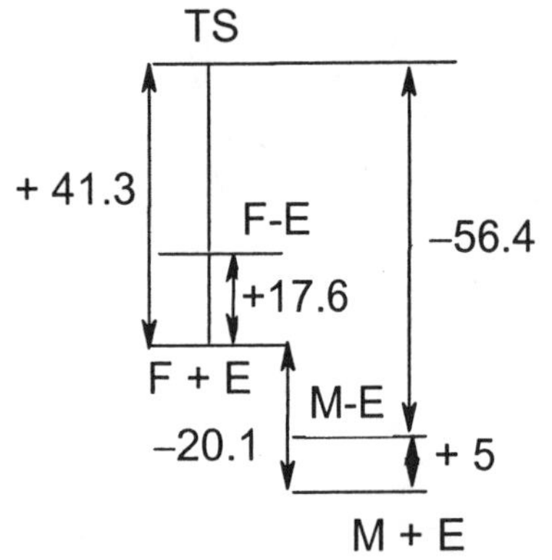

F, E, M and TS represent fumarate ion, enzyme, malate ion and transition state, respectively. The fumarate-fumarate complex is represented as F-E and the malate-fumarate complex is represented as M-E.

(b) The enthalpy of activation of the reverse reaction is 51.4 kJ mol^{-1}.

7.33 The slope of the graph could be estimated as 280 K. When $\Delta^{\ddagger}G$ takes the value of zero, $\Delta^{\ddagger}H$ is approximately $80\ kJ\ mol^{-1}$, which give us an estimation of $\Delta^{\ddagger}G$. The calculation of $\Delta^{\ddagger}G$ at 280 K using several $\Delta^{\ddagger}H$ and $\Delta^{\ddagger}S$ values from the graph confirms that the activation Gibbs energy is the same for different species studied. Read the discussion about this type of graphs in the article: *J. Biosci.*, 2002, 27, 121-126.

7.34 The rate constant at zero ionic strength can be evaluated by using Equation 7.30.

$$\log\ k_{TS} = \log k_{TS}^{0} + 2Az_A z_B I^{1/2}$$

$$\log k_{TS}^{0} = \log 1.55\ -\ 2 \times 0.509 \times (+1) \times (+1) \times \left(0.0241\right)^{1/2}$$

$$k_{TS}^{0} = 1.07\ L^2\ mol^{-3}\ min^{-1}$$

Chapter 8: Complex Biochemical Processes

8.10 (a) Using Fick's first law (Equation 8.1b):

$$J = -D\frac{dc}{dx}$$

$$J = -\left(5.22 \times 10^{-10}\ m^2\ s^{-1}\right) \times \left(-0.1\ mol\ L^{-1}\ m^{-1}\right) \times \left(\frac{10^3\ m^{-3}}{1\ L^{-1}}\right)$$

$$J = 5.2 \times 10^{-8}\ mol\ m^{-2}\ s^{-1}$$

(b) The amount of molecules passing through an area of 5.0 mm^2 is:

$$n = JA\Delta t = 5.2 \times 10^{-8}\ mol\ m^{-2}\ s^{-1} \times 5.0\ mm^2 \times \frac{1\ m^2}{1\times 10^6\ mm^2} \times 60\ s$$

$$n = 1.6 \times 10^{-11}\ mol$$

8.11 The rms distance traveled from starting point in diffusive motion is (Equation 8.3):

$$d = (2Dt)^{1/2}$$

where D is the diffusion constant and t is the elapsed time. The time it takes a molecule to travel a distance d is thus

$$t = d^2/2D$$

Taking diffusion coefficient D of sucrose in water at 25° C from Table 8.1:

(a) For $d = 1$ mm $= 1.0\times10^{-3}$ m

$$t = (1.0\times10^{-3}\ m)^2/\left(2\times(0.522\ 10^{-9}m^2s^{-1})\right) = 958\ s = 16\ minutes$$

(b) For $d = 1$ cm $= 1.0\times10^{-2}$ m

$$t = (1.0\times10^{-2}\ m)^2/\left(2\times(0.522\times10^{-9}m^2s^{-1})\right) = 958\times10^2 s = 27\ hours$$

(c) For $d = 1$ m

$$t = (1.0\ m)^2/\left(2\times(0.522\times10^{-9}m^2s^{-1})\right) = 958\times10^6 s = 30\ years$$

The time grows with the square of distance, becoming very large for macroscopic distances such as 1 m.

8.12 (a) To estimate the diffusion coefficient, we can use the Einstein-Smoluchowski Equation (Equation 8.4).

$$D = \frac{\lambda^2}{2\tau}$$

$$D = \frac{\left(150 \times 10^{-12}\ \text{m}\right)^2}{2 \times 1.8 \times 10^{-12}\ \text{s}}$$

$$D = 6.2 \times 10^{-9}\ \text{m}^2\ \text{s}^{-1}$$

(b)

$$D = \frac{\left(75 \times 10^{-12}\ \text{m}\right)^2}{2 \times 1.8 \times 10^{-12}\ \text{s}}$$

$$D = 1.6 \times 10^{-9}\ \text{m}^2\ \text{s}^{-1}$$

8.13 Following the reasoning in the solution to Exercise 8.11:

$$t = d^2/2D = (1.0\times10^{-6}\ \text{m})^2/\left(2\times(1.0\times10^{-11}\ \text{m}^2\text{s}^{-1})\right) = 0.05\ \text{s} = 50\ \text{ms}$$

This is a time scale similar to many metabolic processes. Thus, diffusion is a fast enough process to transport macromolecules through the cell.

8.14 The time for a lipid to diffuse over a distance in a plasma membrane can be calculated by using Equation 8.3.

$$t = \frac{d^2}{2D}$$

$$t = \frac{\left(10 \times 10^{-9}\ \text{m}\right)^2}{2 \times 1.0 \times 10^{-10}\ \text{m}^2\ \text{s}^{-1}}$$

$$t = 0.5\ \mu\text{s}$$

In a lipid bilayer:

$$t = \frac{d^2}{2D}$$

$$t = \frac{\left(10 \times 10^{-9}\ \text{m}\right)^2}{2 \times 1.0 \times 10^{-9}\ \text{m}^2\ \text{s}^{-1}}$$

$$t = 0.05\ \mu\text{s}$$

8.15 Following the reasoning in the solution to Exercise 8.11, we can write:

$t_1 = d^2/2D_1$: time for ribonuclease to diffuse over distance d

$t_2 = d^2/2D_2$: time for catalase to diffuse over distance d

where D_1 and D_2 are the diffusion coefficients of the proteins.

Given $D \propto M^{-1/2}$, we can write the following equation for the ratio of travel times:

$$t_2/t_1 = (d^2/2D_2)/(d^2/2D_1) = (D_1/D_2) = (M_2/M_1)^{1/2}$$

$$= (250 \text{ kDa}/13.683 \text{ kDa})^{1/2} = 4.27$$

The larger catalase will diffuse about four times more slowly than the smaller ribonuclease, assuming that both proteins are roughly spherical.

8.16 Using the diffusion coefficient of water:

$$t = \frac{d^2}{2D}$$

$$t = \frac{(100 \text{ m})^2}{2 \times 2.26 \times 10^{-9} \text{ m}^2 \text{ s}^{-1}}$$

$$t = 2.21 \times 10^{12} \text{ s}$$

8.17 The rms distance traveled from starting point in diffusive motion over time *t* is (Equation 8.3):

$$d = (2Dt)^{1/2}$$

Treating the diffusion as a random walk, in which a step of length λ is taken every time a period of time τ elapses, we can say that the molecule takes N steps to cover the distance d, with

$$t = N\tau \text{ and } d = n\lambda$$

Thus, the molecule taking N random steps travels n steps away from the origin, on average. Substituting for t and d we obtain the relation between N and n:

$$n\lambda = (2D \times N\tau)^{1/2}$$

solving for N gives

$$N = (n^2\lambda^2)/(2D\tau)$$

We now introduce the Einstein-Smoluchowski equation (Equation 8.4):

$$D = \lambda^2/2\tau$$

Substituting this into our expression for N gives

$$N = n^2 \text{ or } n = N^{1/2}$$

To move a distance of n steps from the origin, we need to take n^2 steps of a one-dimensional random walk.

8.18 The activation energy for the motion of water molecules can be calculated by using Equation 8.7 and taking the logarithm of the ratio of the viscosities at two temperatures:

$$\ln\left(\frac{D_1}{D_2}\right) = \frac{-E_a}{R}\left(\frac{1}{T_1} - \frac{1}{T_2}\right), \text{ but, } D \sim \frac{1}{\eta}, \text{ so}$$

$$\ln\left(\frac{D_1}{D_2}\right) = \ln\left(\frac{\eta_2}{\eta_1}\right) = \frac{-E_a}{R}\left(\frac{1}{T_1} - \frac{1}{T_2}\right)$$

$$\ln \frac{\eta_1}{\eta_2} = \frac{E_a}{R}\left(\frac{1}{T_1} - \frac{1}{T_2}\right)$$

$$\ln \frac{7.982 \times 10^{-4}}{1.0019 \times 10^{-3}} = \frac{E_a}{8.31477\ \text{J K}^{-1}\ \text{mol}^{-1}} \times \left(\frac{1}{303} - \frac{1}{293}\right)\text{K}^{-1}$$

$$E_a = 16.8\ \text{kJ mol}^{-1}$$

8.19 The drift speed s of an ion is (Equation 8.10):

$$s = uE$$

where E is the electric field and u is the ionic mobility.
We can calculate E from the potential difference V and distance d between electrodes

$$E = V/d$$

Thus,

$$s = uE = uV/d = (5.19\times10^{-8}\ \text{m}^2\text{s}^{-1}\text{V}^{-1})(12.0\ \text{V})/(1.00\times10^{-2}\text{m}) = 6.23\times10^{-5}\text{m s}^{-1}$$

This speed is quite small compared to the velocities of random thermal motion. NOTE: the viscosity of water is given in this problem, but is not needed for the solution.

8.20 (a) Exercise 4.45 shows how to get isoelectric points for different types of amino acids.
To determine the charge a molecule of calf thyme histone bears at pH 7, we need to calculate the isoelectric point for each residue by using pKa values from Table 4.6 and the findings from Exercise 4.45.
The composition of one molecule of the protein and the isoelectric point for each residue are shown below:

Amino acid	Number of residues	pI
Aspartic acid	1	pI = ½(pK_{a1} + pK_{a2}) = 2.83
Glutamic acid	1	pI = ½(pK_{a1} + pK_{a2}) = 3.15
Lysine	11	pI = ½(pK_{a2} + pK_{a3}) = 9.91
Arginine	15	pI = ½(pK_{a2} + pK_{a3}) = 10.55
Histidine	2	pI = ½(pK_{a2} + pK_{a3}) = 7.56

Given the protein is made of amino acids, for the most part, with basic sidechains, it will be expected the charge of the protein at pH 7 to be positive. Using the same arguments, the pI of the protein it is expected to be around 10 since the greater percentage of amino acids contributing to the structure of the protein have a pI of around 10.

(b) Following a similar procedure for egg albumin, the composition of one molecule of the protein and the isoelectric point for each residue are shown below:

Aminoacid	Number of residues	pI
Aspartic acid and glutamic acid	51	$pI = ½(pK_{a1} + pK_{a2}) = 2.83$ $pI = ½(pK_{a1} + pK_{a2}) = 3.15$
Lysine	20	$pI = ½(pK_{a2} + pK_{a3}) = 9.91$
Arginine	15	$pI = ½(pK_{a2} + pK_{a3}) = 10.55$
Histidine	7	$pI = ½(pK_{a2} + pK_{a3}) = 7.56$

This case is the opposite of the one discussed in part (a). This protein is made of 55% amino acids with acidic sidechains and 45% amino acids with basic sidechains. Within that 45% we are including a 7% contribution from histine that has a pI close to 7.
This analysis suggests the pI of the protein is less than 7.
(c) Given the pI of the two proteins are expected to be quite different, the isoelectric focusing technique could be a good technique that can be used to separate a mixture of the two proteins.

8.21 Solution using the original data for the K^+ ion.
(a) Taking the mobility of the K^+ ion from Table 8.2, we use the Einstein relation to calculate *D*.

$$D = uRT/zF = (7.62\times10^{-8}\ \text{m}^2\text{s}^{-1}\ \text{V}^{-1})\times(8.314\ \text{J mol}^{-1}\text{K}^{-1})\times(298\ \text{K})/\left((1)(96{,}485\ \text{C mol}^{-1})\right)$$

$$= 1.96\times10^{-9}\,\text{m}^2\text{s}^{-1}$$

Now use the Einstein-Stokes relation (Equation 8.6), linking the diffusion coefficient *D* and hydrodynamic radius *a*:

$$D = kT/6\pi\eta a$$

to find *a*:

$$a = kT/6\pi\eta D$$

$$= (1.381\times10^{-23}\ \text{J K}^{-1})\times(298\ \text{K})/\begin{bmatrix}(6\pi)\times(8.91\times10^{-4}\,\text{kg m}^{-1}\text{s}^{-1})\times\\(1.96\times10^{-9}\text{m}^2\text{s}^{-1})\end{bmatrix}$$

$$= 125\ \text{pm}$$

(b) The ionic radius of K^+ in Table 9.3 is 138 pm. Thus, the hydrodynamic radius of this ion in aqueous solution at room temperature is quite similar to the radius found in ionic crystals.
Solution using data for the Na^+ ion.
(a) Taking the mobility of the Na^+ ion from Table 8.2, we use the Einstein relation to calculate *D*

$$D = uRT/zF$$
$$= (5.19\times10^{-8}\ m^2s^{-1}V^{-1}) \times (8.314\ J\ mol^{-1}K^{-1}) \times (298\ K)/\left[(1)(96{,}485\ C\ mol^{-1})\right]$$
$$= 1.33\times10^{-9} m^2s^{-1}$$

Now use Einstein-Stokes relation (Eq. 8.6) linking diffusion coefficient D and hydrodynamic radius a:

$$D = kT/6\pi\eta a$$

to find a

$$a = kT/6\pi\eta D$$
$$= (1.381\times10^{-23}\ J\ K^{-1})(298\ K)/\ \left[(6\pi)\times(8.91\times10^{-4} kg\ m^{-1}s^{-1})\times(1.33\times10^{-9} m^2s^{-1})\right]$$
$$= 184\ pm$$

(b) The ionic radius of Na^+ in Table 9.3 is 102 pm. Thus, the hydrodynamic radius of this ion in aqueous solution at room temperature is about 80% larger than the radius found in ionic crystals. This effect is explained by the fact that strong electrostatic interactions of Na^+ with water lead to dragging of water molecules by the cation as it migrates through the solution. The volume difference between a sphere with radius a = 184 pm and a sphere with a = 102 pm is about 22×10^6 pm. A water molecule may be approximated by a sphere of radius a = 140 pm, with a volume of about 11×10^6 pm. Thus, the increase in radius of the Na^+ ion is consistent with the dragging of two water molecules.

8.22 The rate of formation of ES is given by:

$$\nu = k_b[ES]$$

We find the concentration of the enzyme-substrate complex by using the pre-equilibrium rate constants and concentrations of enzyme and substrate.
The expression for $K_M = (k_a' + k_b)/k_a$ could become equal to k_a'/k_a if the value of $k_a' \gg k_b$ which implies the assumption of a rapid pre-equilibrium of E, S and ES. Since Michaelis and Menten derived their rate law assuming the mentioned pre-equilibrium, we define the dissociation constant K_d that resembles K_M

$$K_d = k_a'/k_a = \frac{[E][S]}{[ES]}$$

As described in Derivation 8.2, the concentrations of enzyme and the substrate added is equal to $[E]_0 = [E] + [ES]$.
Substitution of this expression into K_d yields:

$$K_d = \frac{([E]_0 - [ES])[S]}{[ES]} = \frac{k_a'}{k_a}$$

Solving for [ES]:

$$[ES] = \frac{[E]_0[S]}{K_d + [S]}$$

Substitution of the concentration of the enzyme-substrate complex into the rate of product formation ($\nu = k_b[ES]$) yields:

$$\nu = \frac{k_b[E]_0[S]}{K_d + [S]}$$

Finally, we divide numerator and denominator of this expression to obtain:

$$\nu = \frac{k_b[E]_0}{1 + K_d/[S]}$$

This expression becomes Equation 8.13 if K_d is identical to K_M. In other words, both rate laws are identical if the assumption of pre-equilibrium is valid.

8.23 The reaction mechanism has four steps, which are described by Equations 8.12a, 8.12b, 8.17a and 8.17b.

We proceed like we did for the Michaelis-Menten mechanism; start by writing the overall reaction rate as the rate of product formation (Equations 8.17a and 8.17b).

$$\nu = k_b[ES] - k_b'[E][P]$$

The next step is to use the steady-state approximation for the central intermediate in our reaction mechanism, the ES complex. We assume that we reach a stage in the reaction when the rates of formation and consumption of ES are equal:

$$0 = d[ES]/dt = k_a[E][S] - k'_a[ES] - k_b[ES] + k'_b[E][P]$$

which we can transform to

$$(k_a' + k_b)[ES] = (k_a[S] + k_b'[P])[E]$$

or

$$[ES] = [E]\,(k_a[S] + k_b'[P])/\,(k_a' + k_b)$$

We now introduce the total enzyme concentration

$$[E]_0 = [E] + [ES]$$

and use our formula for [ES]

$$[E]_0 = [E] + [ES] = [E]\,(\,1 + (k_a[S] + k_b'[P])/\,(k_a' + k_b))$$

$$= [E]\,(\,1 + [S]/K_M + [P]/K_M')$$

where K_M is the Michaelis constant and K_M' is an analogous constant for the reverse reaction, the enzyme-catalyzed formation of S from P:

$$K_M = (k_a' + k_b)/k_a \quad \text{and} \quad K_M' = (k_a' + k_b)/\,k_b'$$

This gives for [ES]

$$[ES] = [E](\,[S]/K_M + [P]/K_M')$$

Now substitute for [E] and [ES] into the rate expression:

$$\nu = k_b[ES] - k_b'[E][P] = k_b(\,[S]/K_M + [P]/K_M') - k_b'[P])[E]$$

$$= [E]_0\,(k_b(\,[S]/K_M + [P]/K_M') - k_b'[P])/(\,1 + [S]/K_M + [P]/K_M')$$

$$= [E]_0\,(k_b \times [S]/K_M - k_a' \times [P]/K_M')/\,(\,1 + [S]/K_M + [P]/K_M')$$

$$= (\nu_{max} \times [S]/K_M - \nu_{max}' \times [P]/K_M')/\,(\,1 + [S]/K_M + [P]/K_M')$$

where we have introduced the maximum rates for the forward and reverse reactions:

$$\nu_{max} = k_b \times [E]_0 \quad \text{and} \quad \nu_{max}' = k_a' \times [E]_0$$

This gives the desired expression.

To test the limiting behavior, do the following:

1) For low product concentrations, [P] ≈ 0, ignore the terms with [P]:

$$v = (v_{max}[S]/K_M)/(1 + [S]/K_M) = v_{max}/(1 + K_M/[S])$$

After substituting $[S] \approx [S]_0$, this gives the Michaelis-Menten mechanism.

2) For low concentrations of both product and substrate, $[S]/K_M << 1$ and [P] << 1, ignore terms with [S] and [P] in denominator.

$$= (v_{max} \times [S]/K_M) - (v'_{max} \times [P]/K'_M)$$

The reaction proceeds in forward or reverse direction, depending on magnitudes of $v_{max} \times [S]/K_M$ and $v'_{max}[P]/K'_M$. If concentration of [P] is much lower than [S], then $v = v_{max}[S]/K_M = (k_b/K_M)[E]_0[S]$.

Again, this is the same result as for Michaelis-Menten.

3) For large concentrations of substrate, $[S]/K_M >> 1$ and low [P], ignore both terms with [P], and the 1 in the denominator:

$$v = v_{max}$$

Forward reaction proceeds with maximum rate.

4) For large concentrations of product, $[P]/K'_M >> 1$ and low [S], ignore both terms with [S], and the 1 in the denominator:

$$v = v'_{max}$$

Reverse reaction proceeds with maximum speed.

8.24 The rate of product formation is $v = k_c[ES']$.

We can obtain the [ES'] by applying the steady-state approximation:

$$\frac{d[ES']}{dt} = k_b[ES] - k_c[ES'] = 0$$

From this, we get: $k_b/k_c[ES] = [ES']$.

Substitution of [ES'] into the rate product formation yields: $v = k_b[ES]$.

We will apply the steady-state approximation to obtain an expression for the concentration of the enzyme-substrate complex ES:

$$\frac{d[ES]}{dt} = k_a[E][S] - k'_a[ES] - k_b[ES] = 0$$

The concentrations of enzyme and substrate added is equal to $[E]_0 = [E] + [ES] + [ES']$, so the expression changes to:

$$\frac{d[ES]}{dt} = k_a([E]_0 - [ES] - [ES'])[S] - k'_a[ES] - k_b[ES] = 0$$

Collecting terms that only contain [ES] and [ES'] on one side of the equation, we get:

$$k_a[E]_0[S] = k_a[ES][S] + k_a[ES'][S] + k'_a[ES] + k_b[ES]$$

Substitution of the expression of $[ES']$:

$$k_a[E]_0[S] = k_a[ES][S] + \frac{k_a k_b}{k_c}[ES][S] + k_a'[ES] + k_b[ES]$$

Then solve for [ES]:

$$[ES] = \frac{k_a[E]_0[S]}{k_a[S] + \frac{k_a k_b}{k_c}[S] + k_a' + k_b}$$

Substitution of [ES] into the rate product formation: $\nu = k_b[ES]$

$$\nu = \frac{k_b k_a[E]_0[S]}{k_a[S] + \frac{k_a k_b}{k_c}[S] + k_a' + k_b}$$

Changing the denominator of the expression:

$$\nu = \frac{k_b k_a[E]_0[S]}{\frac{k_c k_a[S] + k_a k_b[S] + \left(k_a' + k_b\right)k_c}{k_c}}$$

Which can be rearranged to:

$$\nu = \frac{k_b k_a k_c[E]_0[S]}{k_c k_a[S] + k_a k_b[S] + \left(k_a' + k_b\right)k_c}$$

Dividing numerator and denominator by k_a:

$$\nu = \frac{k_b k_c[E]_0[S]}{(k_c + k_b)[S] + \frac{\left(k_a' + k_b\right)k_c}{k_a}}$$

Dividing numerator and denominator by $(k_c + k_b)$:

$$\nu = \frac{k_b k_c/(k_c + k_b)[E]_0[S]}{[S] + \frac{\left(k_a' + k_b\right)k_c}{\left(k_c + k_b\right)k_a}}$$

Dividing numerator and denominator by [S]:

$$\nu = \frac{k_b k_c/(k_c + k_b)[E]_0}{1 + \left[\frac{\left(k_a' + k_b\right)k_c}{\left(k_c + k_b\right)k_a}\right]/[S]}$$

The denominator $k_b k_c/(k_c + k_b)[E]_0$ is identified as ν_{max} and K_M equal to

$$\frac{\left(k_a' + k_b\right)k_c}{\left(k_c + k_b\right)k_a}$$

8.25 The Michaelis-Menten mechanism leads to this expression for reaction rate v

$$v = v_{max}/(1 + K_M/[S]_0)$$

which gives for v_{max}:

$$v_{max} = v(1 + K_M/[S]_0) = (1.15 \text{ mmol L}^{-1}\text{s}^{-1})\left[1+(0.045 \text{ mol L}^{-1})/(0.110 \text{ mol L}^{-1})\right]$$

$$= 1.62 \text{ mmol L}^{-1}\text{s}^{-1}$$

8.26 If $[S]_0 = K_M$, Equation 8.13 changes to:

$$v = \frac{k_b[E]_0}{2} = \frac{v_{max}}{2}$$ (i.e., when the enzyme concentration reach half its initial value).

8.27 Transform the data to $1/v$ and $1/[S]_0$

$1/[S]_0$ / (μmol L^{-1}) 3.145×10^{-2} 2.155×10^{-2} 1.686×10^{-2} 8.439×10^{-3} 4.500×10^{-3}

$1/v$ / (pmol $L^{-1}s^{-1}$) 1.428×10^{-2} 1.029×10^{-2} 8.569×10^{-3} 6.281×10^{-3} 5.141×10^{-3}

and the Michaelis-Menten equation into the Lineweaver-Burk form:

$$1/v = 1/v_{max} + (K_M/v_{max})(1/[S]_0)$$

Plotting $1/v$ vs. $1/[S]_0$ should give a straight line with slope K_M/v_{max} and offset $1/v_{max}$.

Performing a least-squares fit of the data to a straight line (linear regression) yields:

intercept = 3.359×10^{-2} and slope = 0.3354

Thus,

$$v_{max} = 1/\left(3.359\times10^{-2}\right) = 297.7 \text{ (pmol L}^{-1}\text{s}^{-1}\text{)}$$

$$K_M = v_{max} \times (K_M/v_{max}) = (297.7)(0.3354) = 99.8 \text{ (mol L}^{-1}\text{)}$$

The results of the fit have the units implied by the data: $-K_M$ is the same as [S] (μmol L^{-1}) and v_{max} is the same as v (pmol $L^{-1}s^{-1}$).

8.28 We will construct a Lineweaver-Burk plot to obtain kinetic information about the reaction of an ATPase on ATP. According to Equation 8.16, the Lineweaver-Burk plot is a plot of $1/v$ versus $1/[S_0]$. The following table shows $1/v$ and $1/[ATP]$

1/[ATP]/μmol^{-1} L	1.66	1.25	0.71	0.50	0.33
$1/v$/ μmol^{-1} L s	1.23	1.03	0.769	0.680	0.592

The graph is plotted below:

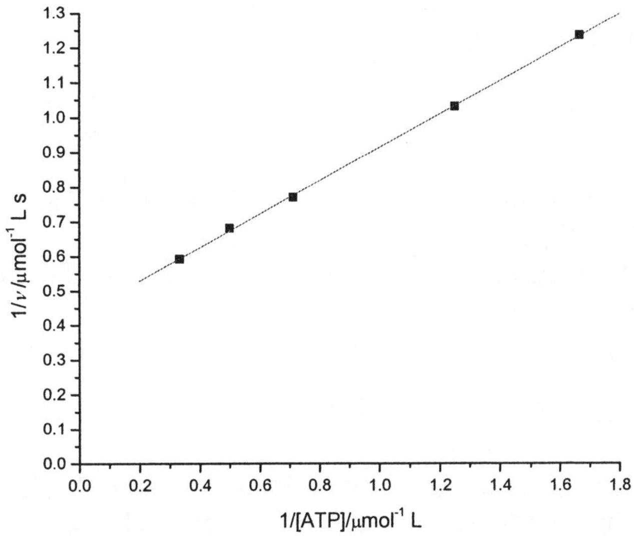

A least-square fit using Origin 7.0 gives a y-intercept at 0.433 and a slope of 0.479.

Using this information and Equation 8.16, it follows that:

$$\nu_{max} / (\mu\text{mol L}^{-1}\,\text{s}^{-1}) = \frac{1}{\text{intercept}} = \frac{1}{0.433} = 2.31$$

and

$$K_M / (\mu\text{mol L}^{-1}) = \frac{\text{slope}}{\text{intercept}} = \frac{0.479}{0.433} = 1.106$$

The turnover number, or catalytic constant, k_{cat} is identified with k_b, according to Equation 8.23. Since the maximum velocity has been determined and the concentration of the enzyme is provided, we can determine k_{cat}:

$$k_{cat} = \frac{\nu_{max}}{[E]_0} = \frac{2.31\ \mu\text{mol L}^{-1}\ \text{s}^{-1}}{0.020\ \mu\text{mol L}^{-1}} = 1.1 \times 10^2\ \text{s}^{-1}$$

The catalytic efficiency of the enzyme, ε, is the ratio k_{cat}/K_M as shown in Equation 8.24. In this case:

$$\varepsilon = \frac{1.1 \times 10^2\ \text{s}^{-1}}{1.106\ \mu\text{mol L}^{-1}} = 99\ \mu\text{mol}^{-1}\ \text{L}\ \text{s}^{-1}$$

8.29 (a) To obtain the Eadie-Hofstee form of the Michaelis-Menten equation start with

$$\nu = \nu_{max}/(1 + K_M/[S]_0) = \nu_{max}\,[S]_0/([S]_0 + K_M)$$

$$\nu/[S]_0 = \nu_{max}/([S]_0 + K_M)$$
$$(\nu/[S]_0)\times([S]_0 + K_M) = \nu_{max}$$
$$\nu + (\nu/[S]_0)K_M = \nu_{max}$$

and finally

$$\nu/[S]_0 = (\nu_{max} - \nu)/K_M$$

(b) Based on part (a), the Eadie-Hofstee plot of $\nu/[S]_0$ vs. ν should give a straight line with a slope of $-1/K_M$ and intercept of ν_{max}/K_M.

(c) Least-squares fit of the transformed data from Exercise 8.27 give the results:

intercept = 3.126 and slope = -1.121×10^{-2}

$$K_M = -1/(-1.121\times10^{-2}) = 89.2 \text{ (mol L}^{-1}\text{)}$$

$$\nu_{max} = K_M(\nu_{max}/K_M) = (89.2)(3.126) = 278.8 \text{ (pmol L}^{-1}\text{s}^{-1}\text{)}$$

These parameters are close to, but not the same as, those obtained in Exercise 8.27. The data contain some "noise" in the form of small random errors, so even if the reaction follows the Michaelis-Menten model exactly, the parameters can only be determined from the data with some finite precision. Different ways of fitting nonlinear model parameters to data will also lead to slightly different parameter values.

8.30 (a) If Equation 8.16 is multiplied by $[S]_0$ we obtain:

$$\frac{[S]_0}{\nu} = \frac{K_M}{\nu_{max}} + \frac{[S]_0}{\nu_{max}}$$

(b) A plot of $[S]_0/\nu$ versus $[S]_0$ is linear with slope of $1/\nu_{max}$, a y-intercept at K_M/ν_{max}, and an x-intercept at $-K_M$.

(c) A least-square fit using Origin 7.0 gives a y-intercept at 0.479 and a slope of 0.433. Using this information and the equation derived in part (b), it follows that:

$$\nu_{max}/(\mu\text{mol L}^{-1}\text{ s}^{-1}) = \frac{1}{\text{slope}} = \frac{1}{0.433} = 2.31$$

and

$$K_M/(\mu\text{mol L}^{-1}) = \frac{\text{int ercept}}{\text{slope}} = \frac{0.479}{0.433} = 1.106$$

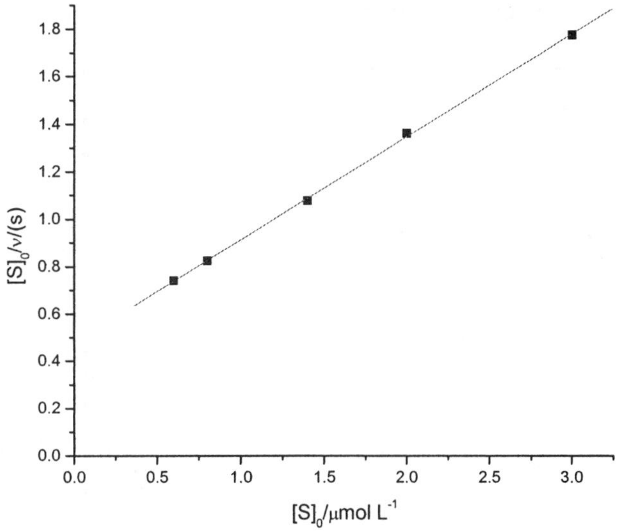

8.31 The figure below shows the effects of cooperativity on enzymatic reaction rate. The function plotted is

$$v/v_{max} = 1/(1 + K'/[S]^n) \qquad \text{with } K' = 1 \text{ nmol L}^{-1}$$

First compare curves with $n = 1$ (standard Michaelis-Menten kinetics) and $n = 2$ (cooperative effect involving two interacting active sites). The $n = 2$ line has a "sigmoidal" or "S" shape—it has a lower slope in the lower and higher range of [S] values, and larger slope in the middle range of [S] values compared to the $n = 1$ line. The $n = 1$ curve is called hyperbolic (i.e., of the type represented by $y = 1/x$).

Second, compare curves with $n = 2$ and $n = 4$, corresponding to different levels of cooperativity. The line for $n = 4$ has a steeper increase in the middle range of [S] values and is more flat in the lower and higher ranges of [S] compared to $n = 2$. Generally, the effect of cooperativity is to introduce an "on/off" type of behavior– low rates at low [S] and high rates at high [S], rather than the gradual increase in rate given by the standard Michaelis-Menten mechanism with $n = 1$.

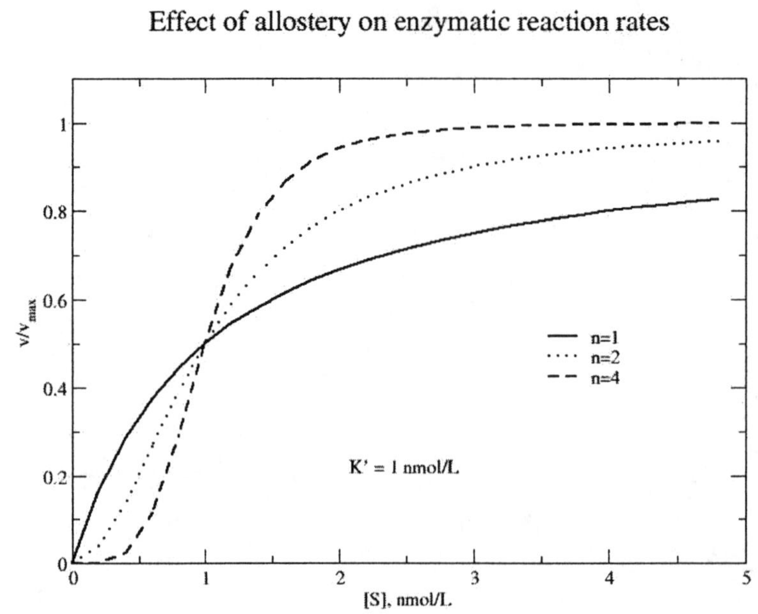

8.32 (a) Taking the reciprocal of the expression:

$$\frac{\nu}{\nu_{max}} = \frac{1}{1 + K'/[S]_0^n}$$

We get:

$$\frac{\nu_{max}}{\nu} - 1 = K'/[S]_0^n$$

Using ν as a common denominator and taking the reciprocal once again yields:

$$\frac{\nu}{\nu_{max} - \nu} = \frac{[S]_0^n}{K'}$$

Finally, by taking logarithm on both sides, we get:

$$\log\frac{\nu}{\nu_{max} - \nu} = n \log [S]_0 - \log K'$$

(b) A plot of $\log \dfrac{v}{v_{max} - v}$ versus $\log [S]_0$ is shown below:

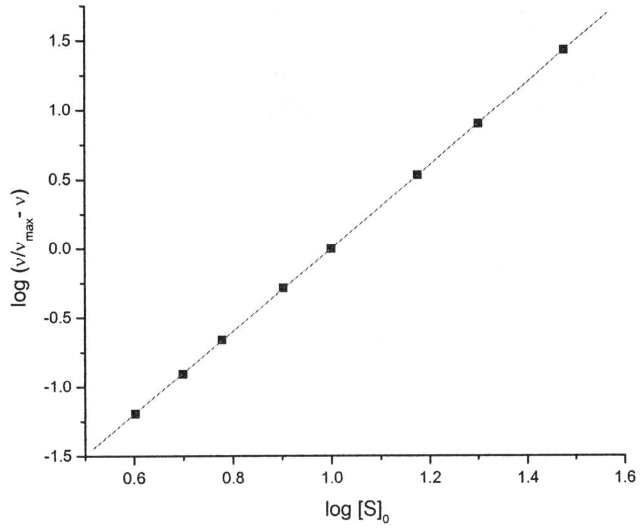

The interaction coefficient n is 3, which is the slope of the graph.

8.33 (a) Solving the equation $v = v_{max}/(1 + K_M/[S]_{10}) = 0.1\, v_{max}$
for $[S]_{10}$, the concentration at which the rate is at 10% maximum, gives
$[S]_{10} = (1/9)\, K_M$
Analogously, solving $v = v_{max}/(1 + K_M/[S]_{90}) = 0.9\, v_{max}$
for $[S]_{90}$, the concentration at which the rate is at 90% maximum, gives
$[S]_{90} = 9\, K_M$.
The ratio is $([S]_{90}/[S]_{10}) = 81$.
(b) Proceeding as in (a), solving the equation $v = v_{max}/(1 + K'/[S]^n_{10}) = 0.1 v_{max}$
for $[S]^n_{10}$ gives $[S]^n_{10} = (1/9)\, K'$.
Solving $v = v_{max}/(1 + K'/[S]^n_{90}) = 0.9\, v_{max}$ for $[S]^n_{90}$, gives $[S]^n_{90} = 9\, K'$.
The ratio is
$([S]^n_{90}/[S]^n_{10}) = ([S]_{90}/[S]_{10})^n = 81$ or equivalently $([S]_{90}/[S]_{10}) = 81^{1/n}$.
(c) From the data in Exercise 8.32, $v_{max} = 4.17$ mol L^{-1} s^{-1}, so we can calculate
$0.1\, v_{max} = 0.42\ \mu\text{mol L}^{-1}\ \text{s}^{-1}$
$0.9\, v_{max} = 3.75\ \mu\text{mol L}^{-1}\ \text{s}^{-1}$
Using the data in the Table in Exercise 8.32, we can estimate
$[S]_{10} \approx 0.50\times 10^{-5}$ mol L^{-1}
$[S]_{90} \approx 2.0\times 10^{-5}$ mol L^{-1}
Thus, $([S]_{90}/[S]_{10}) \approx (2.0/0.50) = 4.0$
The equation for n is $4.0 = 81^{1/n}$
To solve for n, take logarithm of both sides, either log or ln, e.g.

$$\ln 4.0 = (1/n) \ln 81 \quad \text{and} \quad n = (\ln 81)/(\ln 4.0) = 3.2$$

The data indicate that the enzymatic reaction involves cooperativity between 3 active sites.

8.34 A plot of $1/v$ versus 1/[ethanol] (shown below) gives rise to a family of non-parallel lines which is a consistent with a sequential reaction.

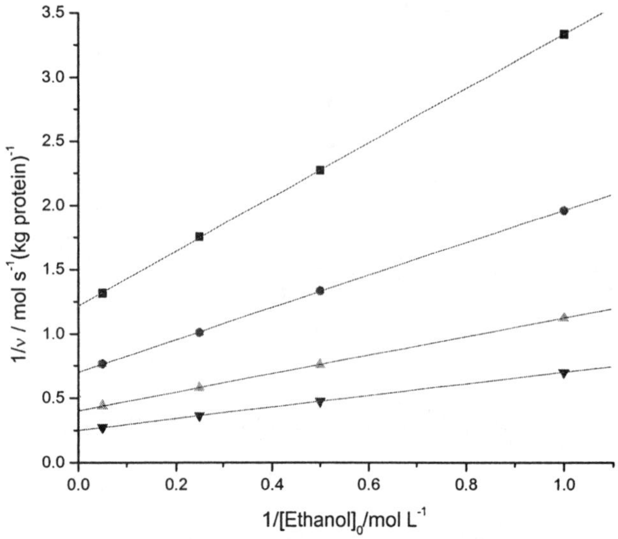

Equation 8.20a can be rearranged to:

$$\frac{1}{v} = \frac{1 + K_M^B/[B]}{v_{max}} + \left(\frac{K_M^A + K_D^A K_M^B/[B]}{v_{max}} \right) \frac{1}{[A]}$$

The slope and y-intercept from the double reciprocal plot are:

$$\text{slope} = \left(\frac{K_M^A + K_D^A K_M^B/[B]}{v_{max}} \right)$$

$$y\text{-intercept} = \frac{1 + K_M^B/[B]}{v_{max}}$$

Evaluation of v_{max} and the appropriate constants can be accomplished by two plots. In one plot, the slopes from the double reciprocal plot are then plotted against 1/[B]. In a second plot, the y-intercepts from the double-reciprocal plot are then plotted against 1/[B].

The values of the slopes and y-intercepts from the double reciprocal plot are shown in the following table:

$[NAD^+]$/mmol L^{-1}	0.050	0.10	0.25	1.0
slope	2.118	1.263	0.722	0.449
y-intercept	1.216	0.698	0.399	0.249

From the intercept plot, v_{max} is the reciprocal of the y-intercept and K_M^B is the slope divided by the y-intercept.

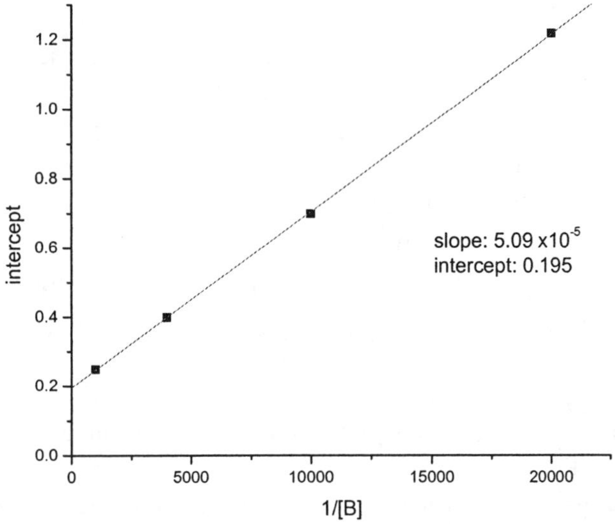

$$v_{max} / (\text{mol L}^{-1}\,\text{s}^{-1}) = \frac{1}{\text{intercept}} = \frac{1}{0.195} = 5.128\text{, and}$$

$$K_M^B / (\text{mol L}^{-1}) = \frac{\text{slope}}{\text{intercept}} = \frac{5.09 \times 10^{-5}}{0.195} = 2.61 \times 10^{-4}$$

From the slope plot, K_M^A is the y-intercept multiplied by v_{max} and K_D^A is the slope divided by the slope of the intercept plot.

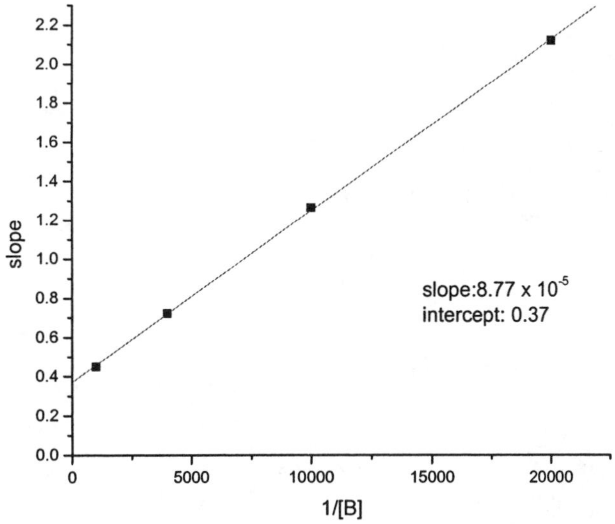

$$K_M^A / (\text{mol L}^{-1}) = \text{intercept} \times v_{max} = 0.37 \times 5.128 = 1.89$$

$$K_D^A / (\text{mol L}^{-1}) = \frac{\text{slope}}{5.09 \times 10^{-5}} = \frac{8.77 \times 10^{-5}}{5.09 \times 10^{-5}} = 1.72$$

8.35 As discussed in the text, the ratio of k_b/K_M is the measure of enzyme catalytic power, with catalytically perfect enzymes achieving values between 10^8 and 10^9 L mol^{-1}s^{-1}. Such enzymatic reactions are called diffusion-controlled, meaning that the chemical step of the reaction occurs essentially as soon as the enzyme and substrate encounter each other in solution.
A reasonable value of K_M is 9.0×10^{-5}mol L^{-1}.

$$k_b / K_M = (1.4\times10^4 \text{s}^{-1})/(9.0\times10^{-5}\text{mol L}^{-1}) = 1.6\times10^8 \text{L mol}^{-1}\text{s}^{-1}$$

Given this value, acetylcholinesterase is a catalytically perfect enzyme.

8.36 We need to construct Lineweaver-Burk plots for different inhibitors and compare them to those in Figure 8.20.
The table below contains the reciprocals of concentration and rates for the three different cases.

1/[CBGP]	$1/v$	1/[CBGP]	$1/v$	1/[CBGP]	$1/v$
0.8	2.51256	0.8	5.81395	0.57143	5.46448
0.26042	1.49477	0.4	3.32226	0.4	4.97512
0.17212	1.16414	0.25	2.90698	0.2	4.329
0.14025	1	0.18182	1.82482	0.1	4.06504

The plot below shows the dependence of rates with concentration for three cases:

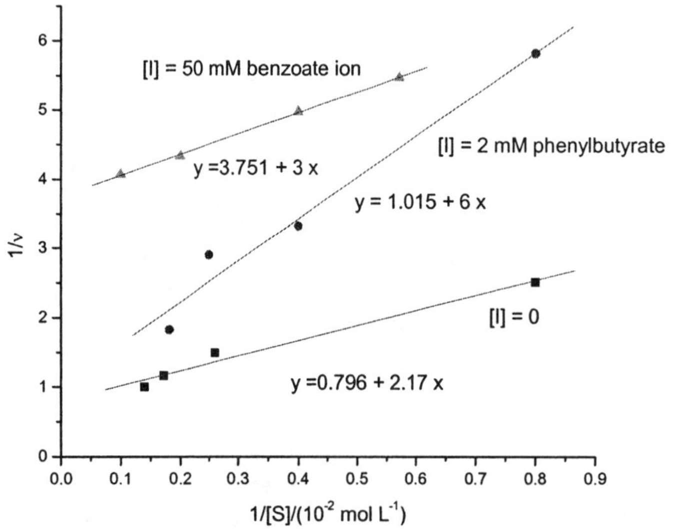

In the case of phenylbutyrate ion, the slopes are very different, which eliminates uncompetitive inhibition. The *y*-intercepts of the plots in the absence of inhibitor and in the presence of 2mM phenylbutyrate are somewhat similar. Given the

uncertainty on the rates values at higher concentration of substrate, in the case of phenylbutyrate we could describe the inhibition as competitive.
In the case of the benzoate ion, the *y*-intercepts of the plots in the absence of inhibitor and in the presence of 50mM benzoate ion are very different, which eliminates competitive inhibition. The slopes are somewhat similar, which indicates uncompetitive inhibition.

8.37 For competitive inhibition

$$v = v_{max}/(1 + \alpha K_M/[S]_0) \quad \text{with} \quad \alpha = (1 + [I]/K_I)$$

where [I] is the inhibitor concentration. In the absence of inhibitor [I] = 0 and $\alpha = 1$:

$$v_0 = v_{max}/(1 + K_M/[S]_0)$$

We are searching for the value of [I] such that

$$v/v_0 = (1 + K_M/[S]_0)/(1 + \alpha K_M/[S]_0) = 0.5$$

Multiplication of both sides by the denominator and solving for α gives

$$\alpha = (0.5 + K_M/[S]_0)/(0.5\,K_M/[S]_0) = (30.5)/15 = 2.03$$

$$\text{using } K_M/[S]_0 = (3.0\times10^{-3}\text{mol L}^{-1})/(1.0\times10^{-4}\text{mol L}^{-1}) = 30$$

Finally,

$$[I] = (\alpha - 1)K_I = (2.03 - 1)(2.0\times10^{-5}\text{mol L}^{-1}) = 2.1\times10^{-5}\text{mol L}^{-1}$$

At low substrate concentrations, the effect of a competitive inhibitor is to slow down the reaction rate by a factor of α.

8.38 (a) If inhibition is taking place due to a high concentration of substrate, the following additional steps should be added to the Michaelis-Menten mechanism:

$$ES + S \longrightarrow ESS \quad v = k_c[ES][S]$$

$$ESS \longrightarrow ES + S \quad v = k_c'[ESS]$$

The rate of product formation is still $v = k_b[ES]$.
We will apply the steady-state approximation to obtain an expression for the concentration of the enzyme-substrate complex ES.

$$\frac{d[ES]}{dt} = k_a[E][S]_0 - k_a'[ES] - k_b[ES] - k_c[ES][S]_0 + k_c'[ESS] = 0$$

If we consider the concentrations of enzyme and substrate added is equal to $[E]_0 = [E] + [ES] + [ESS]$, the expression changes to:

$$\frac{d[ES]}{dt} = k_a([E]_0 - [ES] - [ESS])[S]_0 - k_a'[ES] - k_b[ES] - k_c[ES][S]_0 + k_c'[ESS] = 0$$

Collecting terms that only contain [ES] in one side of the equation, we get:

$$k_a[E]_0[S]_0 = k_a[ES][S]_0 + k_a[ESS][S]_0 + k_a'[ES] + k_b[ES] + k_c[ES][S]_0 - k_c'[ESS]$$

Substitute the expression of [ESS] = [ES][S]/K_1, where K_1 is the equilibrium constant for dissociation of the inhibited enzyme-substrate complex ESS:

$$k_a[E]_0[S] = k_a[ES][S]_0 + \frac{k_a[ES][S]_0^2}{K_1} + k_a'[ES] + k_b[ES] + k_c[ES][S]_0 - \frac{k_c'[ES][S]_0}{K_1}$$

Solve for [ES]:

$$[ES] = \frac{k_a[E]_0[S]_0}{k_a[S]_0 + \frac{k_a[S]_0^2}{K_1} + k_a' + k_b + k_c[S]_0 - \frac{k_c'[S]_0}{K_1}}$$

Substitute [ES] into the rate product formation: $\nu = k_b[ES]$.

$$\nu = \frac{k_b k_a[E]_0[S]_0}{k_a[S]_0 + \frac{k_a[S]_0^2}{K_1} + k_a' + k_b + k_c[S]_0 - \frac{k_c'[S]_0}{K_1}}$$

Which can be rearranged to:

$$\nu = \frac{k_b k_a[E]_0[S]_0}{\left(k_a + \frac{k_a[S]_0}{K_1} + k_c - \frac{k_c'}{K_1}\right)[S]_0 + k_a' + k_b}$$

Divide the numerator and denominator by $k_a[S]$:

$$\nu = \frac{k_b[E]_0}{\left(1 + \frac{[S]_0}{K_1} + \frac{k_c}{k_a} - \frac{k_c'}{k_a K_1}\right) + \frac{k_a' + k_b}{k_a[S]_0}}$$

Since $K_1 = \frac{k_c'}{k_c}$, the expression changes to:

$$\nu = \frac{k_b[E]_0}{\left(1 + \frac{[S]_0}{K_1}\right) + \frac{k_a' + k_b}{k_a[S]_0}}$$

If $k_b[E]_0$ is identified as ν_{max} and $\frac{k_a' + k_b}{k_a}$ as K_M, it follows that:

$$\nu = \frac{\nu_{max}}{1 + \frac{[S]_0}{K_1} + \frac{K_M}{[S]_0}}$$

(b) The graph of $1/v$ against $1/[S]_0$ below assumes a relation between K_I and K_M equal to 10 and 100. In both cases, the linear behavior of Lineweaver-Burk plots is lost at higher substrate concentrations.

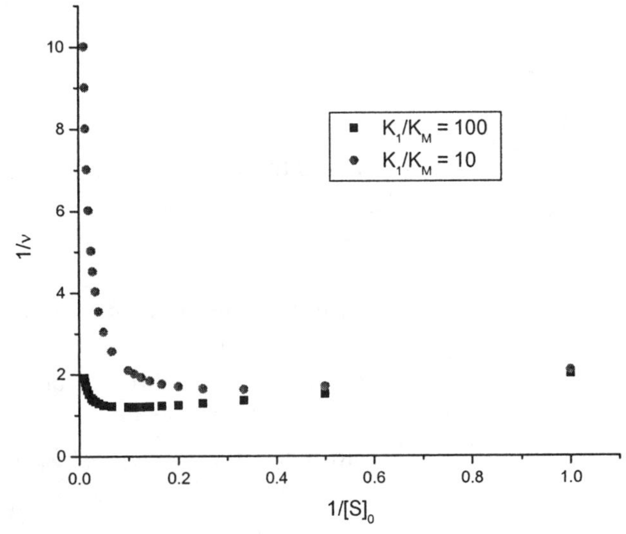

8.39 According to Marcus theory of electron transfer reactions, for a fixed donor-acceptor distance the electron transfer rate depends on the activation energy

$$k_{et} \propto \exp(-\Delta^{\ddagger}G/RT)$$

where

$$\Delta^{\ddagger}G = (\Delta_r G^{-} + \lambda)^2/4\lambda$$

The ratio of the original rate k_{et} to the modified rate k_{et}' may be expressed as

$$k_{et}'/\,k_{et} = (\exp(-\Delta^{\ddagger}G'/RT))/(\exp(-\Delta^{\ddagger}G/RT)) = \exp(-(\Delta^{\ddagger}G' - \Delta^{\ddagger}G)/RT)$$

which we transform to

$$x = \Delta^{\ddagger}G' - \Delta^{\ddagger}G = -RT \times \ln(k_{et}'/\,k_{et})$$

$$= -(8.314\ \text{J mol}^{-1}\text{K}^{-1}) \times (298\ \text{K}) \times \ln(3.33\times10^{6}\text{s}^{-1}/2.02\times10^{5}\text{s}^{-1})$$

$$= -6.94\ \text{kJ mol}^{-1} = -0.072\ \text{eV}$$

Where the variable "x" was introduced to simplify the following equations and the conversion factor 1 eV = 96.48531 kJ/mol was used.

Using equation $\Delta^{\ddagger}G = (\Delta_r G^{-} + \lambda)^2/4\lambda$ from above, we can also write

$$x = \Delta^{\ddagger}G' - \Delta^{\ddagger}G = [((\Delta_r G^{-\prime})^2 + 2\lambda\Delta_r G^{-\prime} + \lambda^2) - ((\Delta_r G^{-})^2 + 2\lambda\Delta_r G^{-} + \lambda^2)]/4\lambda$$

multiplication by 4λ and collecting terms with λ leads to

$$2\lambda\,(\Delta_r G^{-\prime} - \Delta_r G^{-} - 2x) = (\Delta_r G^{-})^2 - (\Delta_r G^{-\prime})^2$$

$$\lambda = [(\Delta_r G^{-})^2 - (\Delta_r G^{-\prime})^2]\,/\,[2\times(\Delta_r G^{-\prime} - \Delta_r G^{-} - 2x)]$$

$$\lambda = [(-0.665)^2 - (-0.975)^2]\,/\,[2\times(-0.975 + 0.665 + 0.144)] = 1.531\ \text{eV}$$

NOTE: Assumed temperature value: 298 K.

8.40 (a) We can find β by using Equation 8.33. We can divide the given rate constants and then take natural logarithm on both sides:

$$\frac{k_{et1}}{k_{et2}} = e^{-\beta(1.11-1.23)}$$

$$\ln\frac{k_{et1}}{k_{et2}} = -\beta(1.11-1.23)$$

Solving for β we get: 16.5 nm^{-1}

(b) If you plot ln k_{et} against r the slope is $-\beta$ and the y-intercept is 30.495. Thus, the estimated value of k_{et} is $4.35 \times 10^2\ s^{-1}$.

8.41 Proceeding analogously as in Example 8.4, we write down the overall electron transfer reaction

azurin(red) + cytochrome(ox) → azurin(ox) + cytochrome(red)

with the two reduction half-reactions

R: cytochrome(ox) + e– → cytochrome(red) $\quad E_R^- = +0.260$ V

L: azurin(ox) + e^- → azurin(red) $\quad E_L^- = +0.304$ V

The standard reaction voltage is thus

$$E^- = E_R^- - E_L^- = (0.260 - 0.304)\ \text{V} = -0.044\ \text{V}$$

and the equilibrium constant for the overall reaction is

$$\ln K = (v \times F \times E^-)/(R \times T)\ .$$

$v = 1$ is the number of electrons transferred in this case.

$\ln K = (1)(96{,}485\ \text{C mol}^{-1})(-0.044\ \text{V})/(8.314\ \text{J mol}^{-1}\text{K}^{-1})(298\ \text{K})$

$= -1.714$

$K = e^{-1.714} = 0.180$

Now use the Marcus cross-relation for the overall electron transfer rate:

$$k_{obs} = (k_{aa}\, k_{cc}\, K)^{1/2}$$

where k_{aa} and k_{cc} are the electron self-transfer rates for azurin and cytochrome c, respectively. Using the given values of k_{obs} and k_{cc}, and the calculated K, we obtain for azurin

$k_{aa} = k_{obs}{}^2/(k_{cc}\, K) = (1.6\times10^3\ \text{L mol}^{-1}\text{s}^{-1})^2/(1.5\times10^2\ \text{L mol}^{-1}\text{s}^{-1})(0.180)$

$= 9.5\times10^4\ \text{L mol}^{-1}\ \text{s}^{-1}$

Chapter 9: The Dynamics of Microscopic Systems

9.9 The energy of an oscillatory motion can only change by discrete packets, or quanta. The quantum is defined as $h\nu$, where h is the Planck constant and ν is the frequency.

(a) $h\nu = (6.626\times10^{-34}\text{J s})(1.0\times10^{15}\text{s}^{-1}) = 6.6\times10^{-19}\text{J} = 4.0\times10^{2}\text{kJ/mol}$

(b) $h\nu = h/T = (6.626\times10^{-34}\text{J s})/(20\times10^{-15}\text{s}) = 3.3\times10^{-20}\text{J} = 20\text{ kJ/mol}$

where $T = 1/\nu$ is the period of the motion

(c) $h\nu = h/T = (6.626\times10^{-34}\text{J s})/(0.5\text{ s}) = 1.3\times10^{-33}\text{J} = 7.8\times10^{-13}\text{kJ/mol}$

The energy conversion factor 1 kJ/mol = 1.66054×10^{-21} J, or 1 J = 6.02214×10^{20} kJ/mol was used.

9.10 (a) We can calculate the average power output of a photodetector by finding out the total energy in Joules per second. The number of photons collected by the photodetector per second is the total energy divided the energy per photon. If we calculate the energy per photon, we can calculate the total energy.

In the case of light-emitting diodes, the energy per photon of 470 nm light is:

$E = h/c\lambda =$

$6.626\times10^{-34}\text{ J s}\times2.998\times10^{8}\text{ m s}^{-1}/4.70\times10^{-7}\text{ m} = 4.23\times10^{-19}\text{ J/photon}$

The total energy is then

$8.0\times10^{7}\text{ photons}\times4.23\times10^{-19}\text{ J photon}^{-1} = 3.4\times10^{-11}\text{ J}$

The average power output is then $3.4\times10^{-11}\text{ J}/3.8\times10^{-3}\text{ s} = 8.9\times10^{-9}\text{ W}$

(b) In the case of lasers, the energy per photon of 780 nm light is:

$E = h/c\lambda =$

$6.626\times10^{-34}\text{ J s}\times2.998\times10^{8}\text{ m s}^{-1}/7.80\times10^{-7}\text{ m} = 2.55\times10^{-19}\text{ J/photon}$

The total energy is then

$8.0\times10^{7}\text{ photons}\times2.55\times10^{-19}\text{ J photon}^{-1} = 2.0\times10^{-11}\text{ J}$

The average power output is then $2.0\times10^{-11}\text{ J}/3.8\times10^{-3}\text{ s} = 5.3\times10^{-9}\text{ W}$

9.11 The de Broglie wavelength λ of an object is related to its momentum *p*.

$\lambda = h/p$ with $p = mv$ (where *m* is the mass and *v* the velocity)

(a) $\lambda = h/p = (6.626\times10^{-34}\text{J s})/\left[(1.0\times10^{-3}\text{kg})\times(1.0\text{ m s}^{-1})\right] = 6.6\times10^{-31}\text{m}$

(b) $\lambda = h/p = (6.626\times10^{-34}\text{J s})/\left[(1.0\times10^{-3}\text{kg})\times(1.0\times10^{8}\text{m s}^{-1})\right] = 6.6\times10^{-39}\text{m}$

(c) $\lambda = h/p = (6.626\times10^{-34}\text{J s})/\left[(6.642\times10^{-27}\text{kg})\times(1000\text{ m s}^{-1})\right] = 1.00\times10^{-10}\text{m}$

The mass of the helium atom is

$m = M/N_A = (4.00\times\text{g mol}^{-1})/(6.022\times10^{23}\text{mol}^{-1}) = 6.642\times10^{-24}\text{g}$

(d) $\lambda = h/p = (6.626\times10^{-34}\text{J s})/\left[(60\text{ kg})\times(2\text{ m s}^{-1})\right] = 6\times10^{-36}\text{m}$

(e) $\lambda = h/p = \infty$

In (a), (b), and (d), the wavelengths are too small to measure and are much smaller than typical size of the body in question. In (c), the wavelength is comparable to the size of an atom. In (e), the wavelength is too large to measure.

9.12 We can calculate the linear momentum per photon by using Equation 9.3.

(a) $p = h/\lambda = \dfrac{6.626\times10^{-34}\text{ J s}}{600\times10^{-9}\text{ m}} = 1.10\times10^{-27}\text{ kg m s}^{-1}$

Energy per photon of 600 nm radiation is:

$E = h/c\lambda =$

$6.626\times10^{-34}\text{ J s}\times2.998\times10^{8}\text{ m s}^{-1}/6.00\times10^{-7}\text{ m} = 3.31\times10^{-19}\text{ J/photon}$

The energy per mole is then:

$E = 3.31\times10^{-19}\text{ J/photon}\times6.022\times10^{23}\text{ photons mol}^{-1} = 1.99\times10^{5}\text{ J}$

(b) $p = h/\lambda = \dfrac{6.626\times10^{-34}\text{ J s}}{550\times10^{-9}\text{ m}} = 1.20\times10^{-27}\text{ kg m s}^{-1}$

Energy per photon of 550 nm radiation is:

$E = h/c\lambda =$

$6.626\times10^{-34}\text{ J s}\times2.998\times10^{8}\text{ m s}^{-1}/5.50\times10^{-7}\text{ m} = 3.61\times10^{-19}\text{ J/photon}$

The energy per mole is then:

$E = 3.61\times10^{-19}\text{ J/photon}\times6.022\times10^{23}\text{ photons mol}^{-1} = 2.17\times10^{5}\text{ J}$

(c) $p = h/\lambda = \dfrac{6.626\times10^{-34}\text{ J s}}{400\times10^{-9}\text{ m}} = 1.66\times10^{-27}\text{ kg m s}^{-1}$

Energy per photon of 400 nm radiation is:

$E = h/c\lambda =$

$6.626\times10^{-34}\text{ J s}\times2.998\times10^{8}\text{ m s}^{-1}/4.00\times10^{-7}\text{ m} = 4.96\times10^{-19}\text{ J/photon}$

The energy per mole is then:

$E = 4.96\times10^{-19}\text{ J/photon}\times6.022\times10^{23}\text{ photons mol}^{-1} = 2.98\times10^{5}\text{ J}$

(d) $p = h/\lambda = \dfrac{6.626 \times 10^{-34}\ \text{J s}}{200 \times 10^{-9}\ \text{m}} = 3.31 \times 10^{-27}\ \text{kg m s}^{-1}$

Energy per photon of 200 nm radiation is:

$E = h/c\lambda =$

$6.626 \times 10^{-34}\ \text{J s} \times 2.998 \times 10^{8}\ \text{m s}^{-1} / 2.00 \times 10^{-7}\ \text{m} = 9.93 \times 10^{-19}\ \text{J/photon}$

The energy per mole is then:

$E = 9.93 \times 10^{-19}\ \text{J/photon} \times 6.022 \times 10^{23}\ \text{photons mol}^{-1} = 5.98 \times 10^{5}\ \text{J}$

(e) $p = h/\lambda = \dfrac{6.626 \times 10^{-34}\ \text{J s}}{150 \times 10^{-12}\ \text{m}} = 4.41 \times 10^{-34}\ \text{kg m s}^{-1}$

Energy per photon of 150 pm radiation is:

$E = h/c\lambda =$

$6.626 \times 10^{-34}\ \text{J s} \times 2.998 \times 10^{8}\ \text{m s}^{-1} / 1.50 \times 10^{-10}\ \text{m} = 1.32 \times 10^{-15}\ \text{J/photon}$

The energy per mole is then:

$E = 1.32 \times 10^{-15}\ \text{J/photon} \times 6.022 \times 10^{23}\ \text{photons mol}^{-1} = 7.95 \times 10^{8}\ \text{J}$

(f) $p = h/\lambda = \dfrac{6.626 \times 10^{-34}\ \text{J s}}{0.010\ \text{m}} = 6.6 \times 10^{-32}\ \text{kg m s}^{-1}$

Energy per photon of 0.01 m radiation is:

$E = h/c\lambda =$

$6.626 \times 10^{-34}\ \text{J s} \times 2.998 \times 10^{8}\ \text{m s}^{-1} / 0.01\ \text{m} = 1.98 \times 10^{-23}\ \text{J/photon}$

The energy per mole is then:

$E = 1.98 \times 10^{-23}\ \text{J/photon} \times 6.022 \times 10^{23}\ \text{photons mol}^{-1} = 11.9\ \text{J}$

9.13 (a) For the relativistic electron, the wavelength is

$\lambda = h / [2m_e eV(1 + eV/(2m_e c^2))]^{1/2}$

where the classical momentum is $(2m_e eV)^{1/2}$ (see b).

$(2m_e eV)^{1/2} = [2(9.109\times10^{-31}\text{kg})(1.602\times10^{-19}\text{C})(50\times10^{3}\text{V})]^{1/2}$

$= 1.2\times10^{-22}\text{kg m s}^{-1}$

The relativistic correction is

$eV/(2m_e c^2) =$

$(1.602\times10^{-19}\text{C})\times(50\times10^{3}\text{V})/\left[2\times(9.109\times10^{-31}\text{kg})\times(2.998\times10^{8}\text{m s}^{-1})^2\right]$

$= 0.049$

Finally,

$\lambda = (6.626\times10^{-34}\text{J s})/\left[(1.2\times10^{-22}\text{kg m s}^{-1})(1+0.049)^{1/2}\right] = 5.4\times10^{-12}\text{m}$

(b) Classically, the electron gains kinetic energy $p^2/2m_e$ from electrical work eV, so,

$$p^2/2m_e = eV \quad \text{or} \quad p = (2m_e eV)^{1/2}$$
$$\lambda = h/p = h/(2m_e eV)^{1/2} =$$
$$(6.626\times10^{-34}\text{J s})/(1.2\times10^{-22}\text{kg m s}^{-1}) = 5.5\times10^{-12}\text{m}$$

Where the value of $(2m_e eV)^{1/2}$ calculated in (a) was used.
Comparing results of (a) and (b) shows that in this case, the relativistic correction is small.

9.14 (a) Since force is the rate of change of momentum, we will calculate the force exerted by a beam with a wavelength of 650 nm, assuming the power of the laser is 1.0 kW.
First, we calculate the energy per photon of 650 nm radiation:
$E = h/c\lambda =$

6.626×10^{-34} J s $\times$ 2.998×10^{8} m s^{-1} / 6.50×10^{-7} m = 3.06×10^{-19} J/photon

Assuming the power of the laser is 1000 W, or 1000 J s^{-1}, the total number of photons per second is:

$$\text{Number of photons: } \frac{1000 \text{ J s}^{-1}}{3.06 \times 10^{-19} \text{ J photon}^{-1}} = 3.27 \times 10^{21} \text{ photon s}^{-1}$$

We then calculate the change of linear momentum per photon and per second by using Equation 9.3.

$$F = \frac{m \times v}{t} = \frac{h}{\lambda t} = \frac{6.626 \times 10^{-34} \text{ J s}}{650 \times 10^{-9} \text{ m} \times 1 \text{ s}} = 1.02 \times 10^{-27} \text{ kg m s}^{-2}$$

Thus, the force exerted by 3.27×10^{21} photons per second is:

$F = 1.02 \times 10^{-27}$ kg m s^{-2} $\times$ 3.27×10^{21} photons

$F = 3.34 \times 10^{-6}$ kg m s^{-2}

$$\text{(b) } p = F/A = \frac{3.34 \times 10^{-6} \text{ kg m s}^{-2}}{(1000 \text{ m})^2} = 3.34 \times 10^{-12} \text{ Pa}$$

$$\text{(c) time} = \frac{\text{speed}}{\text{(force/mass)}} = \frac{1.0 \text{ m s}^{-1}}{3.34 \times 10^{-6} \text{ kg m s}^{-2}/1.0 \text{ kg}} = 3.0 \times 10^{5} \text{ s},$$

or 83 h.

9.15 The Heisenberg uncertainty principle states that for simultaneous measurements of position and momentum along axis x, the uncertainty in position Δx and the uncertainty in momentum Δp_x are related by

$$\Delta x\, \Delta p_x \geq h/2.$$

In our case,

$$p_x = mv_x = (1.66\times10^{-27}\text{kg})(350\times10^{3}\text{m s}^{-1}) = 5.81\times10^{-22}\text{kg m s}^{-1}$$

The uncertainty of measuring momentum is 0.01% of its value

$$\Delta p_x = (1.0\times10^{-4})(\,5.81\times10^{-22}\text{kg m s}^{-1}) = 5.81\times10^{-26}\text{kg m s}^{-1}$$

Thus,

$$\Delta x \geq h/2\Delta p_x = (6.626\times10^{-34}\text{J s}) \,/\, \left[2\times(5.81\ 10^{-26}\text{kg m s}^{-1})\right] = 5.7\text{ nm}$$

The lowest possible value of position measurement is about 6 nm. Uncertainties of this magnitude must be tolerated, as they cannot be avoided.

9.16 The minimum uncertainty in the electron's position is 100 pm. The uncertainty in the electron's speed can be obtained by first calculating Δp.

$$\Delta p \geq \hbar/2\,\Delta x = \frac{1.0546\times10^{-34}\text{ J s}}{2\times100\times10^{-12}\text{ m}} = 5.27\times10^{-25}\text{ kg m s}^{-1}$$

The uncertainty in the electron's speed is then:

$$\Delta v = \Delta p/m = \frac{5.27\times10^{-25}\text{ kg m s}^{-1}}{9.11\times10^{-31}\text{ kg}} = 5.78\times10^{5}\text{ m s}^{-1}$$

9.17 In accord with the Born interpretation of the wavefunction ψ, the probability of finding the electron between $x=a$ and $x=b$ is

$$P(a\leq x\leq b) = \int_a^b \psi^2(x)\text{d}x$$

If the variation of ψ^2 between a and b is small, we can assume that the function remains constant in the whole range and write

$$P(a\leq x\leq b) \approx (\psi_0^2)\times(b-a)$$

where ψ^2_0 is the value of ψ^2 at the center of the interval, $x_0 = (a+b)/2$. Geometrically, we are assuming that the area under the ψ^2 curve between $x=a$ and $x=b$ is equal to the area of the rectangle of width $b-a$ and height ψ^2_0.

(a) $\text{P}(0.1\text{ nm}\leq x\leq 0.2\text{ nm}) \approx \psi^2(\text{x}=0.15\text{ nm})\times(0.2\text{ nm}-0.1\text{ nm})$

$$(1.77\times10^{-3}\text{nm}^{-1})(0.1\text{ nm}) = 1.8\times10^{-4}$$

(b) $P(4.9\text{ nm}\leq x\leq 5.2\text{ nm}) \approx \psi^2(x=5.05\text{ nm})\times(5.2\text{ nm}-4.9\text{ nm})$

$$(1.97\times10^{-4}\text{nm}^{-1})(0.3\text{ nm}) = 5.9\times10^{-5}\text{nm}$$

with $\psi(x) = (2/L)^{1/2}\sin(2\pi x/L)$ and $L = 10$ nm.

9.18 The probability of finding the electron is given by the integral:

$$P = \int_{x_1}^{x_2} \psi^2 \mathrm{d}x = \frac{2}{L}\int_{x_1}^{x_2} \sin^2\left(\frac{2\pi x}{L}\right) dx$$

In Derivation 9.1, the required indefinite integral is:

$$\int \sin^2 a\,x\,dx = 1/2\ x - \frac{\sin 2a\,x}{4a}$$

In our case, a = 2π/L = 0.63.

Evaluation of the integral between two limits is then:

$$\frac{2}{L}\left[\left(1/2\ x - \frac{\sin 2a\,x}{4a}\right) - \left(1/2\ x - \frac{\sin 2a\,x}{4a}\right)\right]_{x_1}^{x_2}$$

(a) The probability between $x = 0.1$ nm and $x = 0.2$ nm is:

$$\frac{2}{10}(0.0982 - 0.0491) = 9.82 \times 10^{-3}$$

(b) The probability between $x = 4.9$ nm and $x = 5.2$ nm is:

$$\frac{2}{10}(2.554 - 2.407) = 2.94 \times 10^{-2}$$

The percentage errors in the procedure used in Exercise 9.17 are 98 and 99%, respectively.

9.19 Using the integral

$$\int_a^b \sin^2(kx)\mathrm{d}x = \frac{b-a}{2} - \frac{\sin(2kb) - \sin(2ka)}{4k}$$

with $k = \pi/L$ gives

$$\int_a^b \sin^2(\pi x/L)\mathrm{d}x = \frac{b-a}{2} - \frac{L}{4\pi}(\sin(2\pi b/L) - \sin(2\pi a/L))$$

(a)

$$P(0 \le x \le L/3) = \frac{2}{L}\int_0^{L/3} \sin^2(\pi x/L)\mathrm{d}x = \frac{2}{L}\left(\frac{L/3-0}{2} - \frac{L}{4\pi}(\sin(2\pi/3) - \sin(0))\right)$$

$$= \frac{2}{L}\left(\frac{L}{6} - \frac{L}{4\pi}0.866025\right) = \frac{1}{3} - \frac{0.866025}{2\pi} = 0.1955$$

(b)

$$P(L/3 \le x \le 2L/3) = \frac{2}{L}\int_{L/3}^{2L/3} \sin^2(\pi x/L)\mathrm{d}x = \frac{2}{L}\left(\frac{(2L/3)-(L/3)}{2} - \frac{L}{4\pi}(\sin(4\pi/3) - \sin(2\pi/3))\right)$$

$$= \frac{2}{L}\left(\frac{L}{6} - \frac{L}{4\pi}(-1.732051)\right) = \frac{1}{3} + \frac{1.732051}{2\pi} = 0.6090$$

(c)

$$P(2L/3 \le x \le L) = \frac{2}{L}\int_{2L/3}^{L} \sin^2(\pi x/L)\mathrm{d}x = \frac{2}{L}\left(\frac{L-(2L/3)}{2} - \frac{L}{4\pi}\left(\sin(2\pi) - \sin(4\pi/3)\right)\right)$$

$$= \frac{2}{L}\left(\frac{L}{6} - \frac{L}{4\pi}0.866025\right) = \frac{1}{3} - \frac{0.866025}{2\pi} = 0.1955$$

Check: 0.1955 + 0.6090 + 0.1955 = 1.000—the particle is certainly somewhere in the box.

9.20 To normalize the wavefunction, we need to evaluate the following integral:

$$\int_0^L \mathrm{A}^2 dx = 1$$

$$\mathrm{A}^2 \int_0^L dx = 1$$

$$\mathrm{A}^2\ \mathrm{L} = 1$$

$$\mathrm{A} = \left(\frac{1}{\mathrm{L}}\right)^{1/2}$$

The normalized wavefunction is then: $\psi = \left(\frac{1}{\mathrm{L}}\right)^{1/2}$

9.21 (a) The length of a chain of 12 atoms (11 carbons and 1 oxygen) is
$\mathrm{L} = 11 \times 140\ \mathrm{pm} = 1540\ \mathrm{pm} = 1.54\ \mathrm{nm}$
We assume that electrons of retinal behave like particles in a box of length L. Their allowed energies are then:

$$E_n = n^2h^2/8m_eL^2$$

As stated, the ground state corresponds to all levels up to n=6 occupied by two electrons each (with opposite spins). The first excited state will then correspond to promotion of one electron from energy level n=6 to energy level n=7, while all other electrons remain in their initial states. The energy change of the molecular transition from ground to first excited state may be calculated as the change in electronic energy.

$$\Delta E = E_7 - E_6 = \left(7^2h^2/8m_eL^2\right) - \left(6^2h^2/8m_eL^2\right) = 13h^2/8m_eL^2$$

$$= 13 \times (6.626\ 10^{-34}\mathrm{J\ s})^2 / \left[8 \times (9.109\ 10^{-31}\mathrm{kg}) \times (1.54 \times 10^{-9}\mathrm{m})^2\right] = 3.3\ 10^{-19}\ \mathrm{J}$$

(b) The transition from ground to excited state will occur after the absorption of light having energy $E_l = h\nu$, which matches the energy change in the molecule.

$$h\nu = \Delta E$$

$$\nu = \Delta E/h = (3.3\times10^{-19}\text{J})/(6.626\times10^{-34}\text{J s}) = 5.0\times10^{14}\text{s}^{-1}$$

The wavelength of the absorbed light will be

$$\lambda = c/\nu = (2.998\times10^{8}\text{m s}^{-1})/(5.0\times10^{14}\text{s}^{-1}) = 600\text{ nm}$$

This corresponds to visible light of orange color.

9.22 A ratio of tunneling rates shows how the rates change with distance:

$$\frac{T_1}{T_2} = e^{-2\times 7\text{ nm}^{-1}(1.0\text{ nm} - 2.0\text{ nm})}$$

$$\frac{T_1}{T_2} = 1.2\times10^{6}$$

$$\frac{T_1}{T_2} = \frac{e^{-2\kappa L_1}}{e^{-2\kappa L_2}} = e^{-2\kappa(L_1 - L_2)}$$

The tunneling rates will increase a million-fold when the distance is halved.

9.23 (a) The tunneling rate for width $d = 750$ pm is

$$\nu = Ae^{-d/l} = (5.0\times10^{14}\text{s}^{-1})e^{(-750/70)} = 1.1\times10^{10}\text{s}^{-1}$$

(b) The tunneling rate for width $d = 850$ pm is

$$\nu' = Ae^{-d/l} = (5.0\times10^{14}\text{s}^{-1})e^{(-850/70)} = 2.7\times10^{9}\text{s}^{-1}$$

The ratio of rates: $\nu/\nu' = (1.1\times10^{10}\text{s}^{-1})/(2.7\times10^{9}\text{s}^{-1}) = 4$

The current should be reduced by a factor of 4.

9.24 a) Origin 7.0 can be used to plot wavefunctions and the corresponding probability densities.

The wavefunction $\psi_{1,1}$:

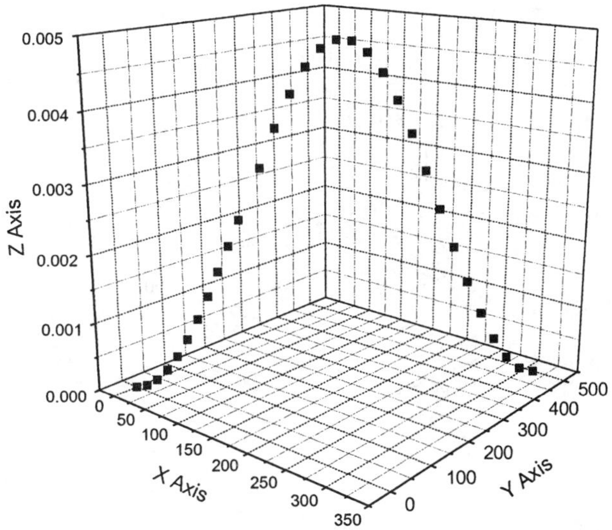

And the corresponding probability densities:

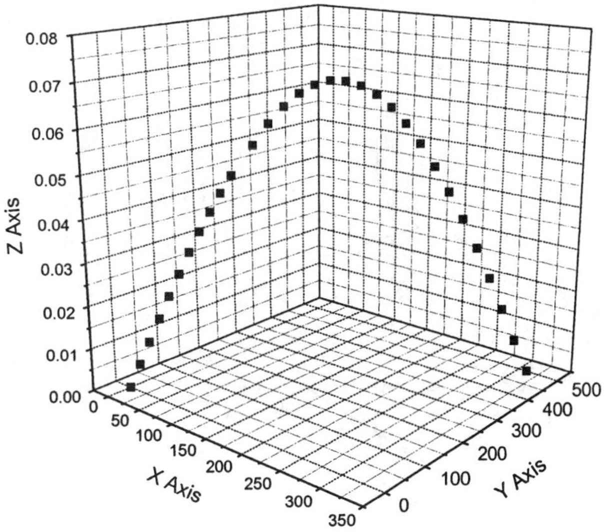

The wavefunction $\psi_{1,2}$:

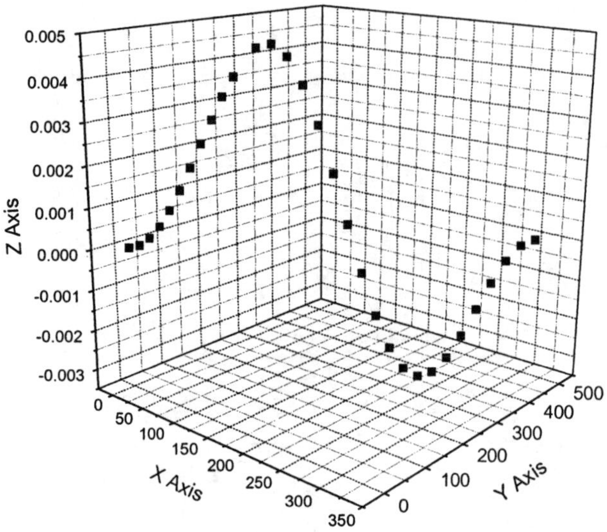

And the corresponding probability densities

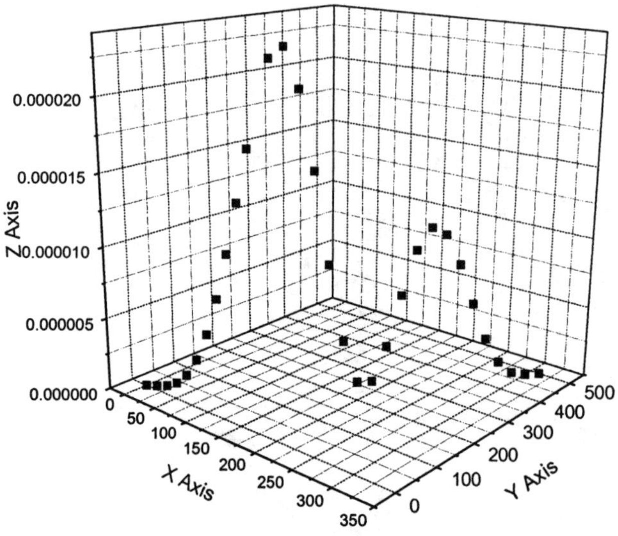

The wavefunction $\psi_{2,1}$:

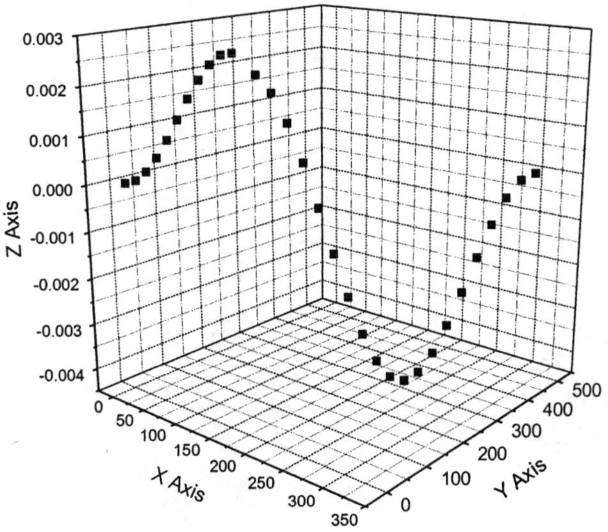

And the corresponding probability densities:

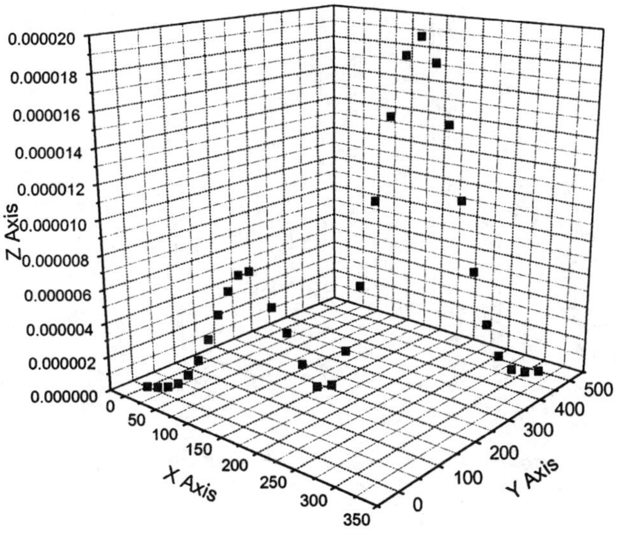

The wavefunction $\psi_{2,2}$:

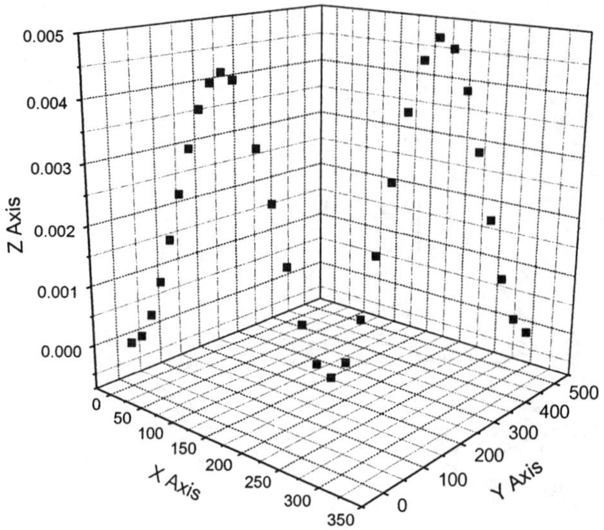

And the corresponding probability densities:

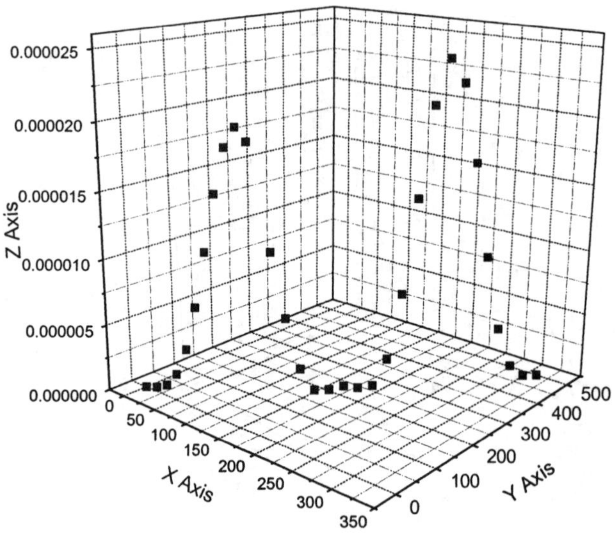

(b) and (c) The first fifteen energy levels of a particle in a rectangular box are:

n1	n2	L1	L2	E ($h^2/8m\ 10^{19}$)
1	1	2.80E-10	4.50E-10	1.769
1	2	2.80E-10	4.50E-10	3.251
2	1	2.80E-10	4.50E-10	5.596
1	3	2.80E-10	4.50E-10	5.720
2	**2**	**2.80E-10**	**4.50E-10**	**7.077**
1	4	2.80E-10	4.50E-10	9.177
2	3	2.80E-10	4.50E-10	9.546
3	1	2.80E-10	4.50E-10	11.973
2	4	2.80E-10	4.50E-10	13.003
3	2	2.80E-10	4.50E-10	13.455
3	3	2.80E-10	4.50E-10	15.924
3	4	2.80E-10	4.50E-10	19.381
4	1	2.80E-10	4.50E-10	20.902
4	2	2.80E-10	4.50E-10	22.383
4	3	2.80E-10	4.50E-10	24.853

The 5th level is in bold, which would be the "highest occupied level".
In the calculations, L_1 was designated as equal to 280 pm and L_2 as 450 pm.
The energy of an electron in the highest occupied level corresponds to E_{22} since there are a total of 10 electrons in the conjugated system and each level is occupied by two electrons.

$$E_{2,2} = (7.077\times10^{19}\,\text{m}^{-2})\times\frac{(6.626\times10^{-34})^2}{8\times(9.110\times10^{-31}\,\text{kg})} = 4.26\times10^{-18}\,\text{J}$$

(d) The change in the energy between the highest occupied and the lowest unoccupied levels corresponds to $E_{14} - E_{22}$

$$\Delta E = E_{1,4} - E_{2,2} = (9.17\text{-}7.07\times10^{19})\times\frac{(6.626\times10^{-34})^2}{8\times(9.110\times10^{-31}\,\text{kg})} = 1.26\times10^{-18}\,\text{J}$$

$$\nu = \Delta E/h = (1.26\times10^{-18}\,\text{J})/(6.626\times10^{-34}\,\text{J s}) = 1.90\times10^{15}\text{s}^{-1}$$

9.25 (a)

(i) For a particle of mass m moving on a circle of radius d, the moment of inertia is $I = md^2 = (1.66\times10^{-27}\text{kg})\times(161\times10^{-12}\text{m})^2 = 4.30\times10^{-47}\text{kg m}^2$

(ii) The allowed energy levels for a planar rotor are

$E_n = n^2\hbar^2/2I \quad n=0, \pm1, \pm2, \ldots \qquad$ with $\hbar = h/2\pi$

The molecule can absorb radiation with energy $E_l = h\nu$ matching the energy difference between allowed energy levels

$$E_l = h\nu = hc/\lambda = \text{E}_\text{n} - E_m$$

Radiation with the greatest wavelength will be absorbed when the transition is between the closest lying energy levels, i.e between n=0 and either n=−1 or n=+1,

$$\text{hc}/\lambda = E_{\pm1} - E_0 = \hbar^2/2I - 0 = \hbar^2/2I$$

$\lambda = 2Ihc/\hbar^2 = (8\pi^2 Ic) / h$ =

$8\pi^2 \times (4.30\times10^{-47}\text{kg m}^2)\times(2.998\times10^8\text{m s}^{-1})/(6.626\times10^{-34}\text{J s})$

$= 1.5\times10^{-3}\text{m} = 1.5 \text{ mm}$

This is in the microwave range, typical for rotational spectra.

(b)

(i) The vibrational frequency of an oscillator is related to mass m and force constant k by

$$\nu = \frac{1}{2\pi}\sqrt{\frac{k}{m}} = \frac{1}{2\pi}\sqrt{\frac{3.14 \text{ N m}^{-1}}{1.66\times10^{-27}\text{kg}}} = 6.92\times10^{13}\text{s}^{-1}$$

Classically, the molecule will vibrate with a frequency about $7\times10^{13}\text{s}^{-1}$.

(ii) The allowed energy levels for the quantum harmonic oscillator are

$E_n = h\nu(n + 1/2) \qquad n = 0,1,2,\ldots$

The molecule can absorb radiation with energy $E_l = h\nu_l = hc/\lambda$ matching the energy difference between allowed energy levels. The simplest kind of transition is from the ground state, n=0, to the first excited state, n=1. Then

$hc/\lambda = E_1 - E_0 = h\nu$

Remember, λ is the wavelength of light, and ν is the oscillator frequency.

$\lambda = hc/h\nu = c/\nu =$

$(2.998\times10^8\text{m s}^{-1})/(6.92\times10^{13}\text{s}^{-1}) = 4.33\times10^{-6}\text{m} = 4.33\ \mu\text{m}$

This is in the infrared range, typical for vibrational spectra.

The frequency of absorbed light will be $\nu_l = c/\lambda = \nu$, equal to the vibrational frequency; this is called resonance.

(c) When H in HI is replaced by deuterium D, the mass *m* of the moving atom will roughly double, while the force constant *k* will remain the same (*k* depends on the electronic structure of the molecule, which is determined by the number of electrons and protons, not on the mass of the nuclei). Thus, the frequency will be decreased by a factor of $2^{-1/2}$.

$\nu_{\text{DBr}} = 2^{-1/2} \times \nu_{\text{HBr}}$

9.26 (a) The energy of an electron in the highest occupied level corresponds to $E_{\pm5}$, since there are a total of 20 electrons in the conjugated system and each level is occupied by two electrons. The $m_l = 0, \pm1, \pm2, \pm3, \pm4,$ and ±5, levels are occupied. Notice that each energy level, other than the one with $m_l = 0$ is doubly degenerate.

$$E_{\pm5} = \frac{5^2 \times (1.054 \times 10^{-34} \text{ J s})^2}{2 \times 9.110 \times 10^{-31} \text{ kg} \times (4.40 \times 10^{-10} \text{ m})^2}$$

$E_{\pm5} = 7.87 \times 10^{-19}$ J

From Equation 9.14, the angular momentum of an electron in the $m_l = \pm5$ level is:

$J_z = m_\ell\hbar = \pm5(1.054\times10^{-34}) = \pm5.27\times10^{-34}\ J\text{sec}$

(b) The change in the energy between the highest occupied and the lowest unoccupied levels corresponds to $E_{\pm 6} - E_{\pm 5}$

$$\Delta E = E_{\pm 6} - E_{\pm 5}$$

$$\Delta E = (36 - 25)\frac{(1.054 \times 10^{-34}\ \text{J s})^2}{2 \times 9.110 \times 10^{-31}\ \text{kg} \times (4.40 \times 10^{-10}\ \text{m})^2}$$

$$\Delta E = 3.46 \times 10^{-19}\ \text{J}$$

Using Equation 9.9, the frequency of radiation that can induce a transition between the $m_l = \pm 5$ and the $m_l = \pm 6$ levels is:

$$\nu = \frac{\Delta E}{h} = \frac{3.46 \times 10^{-19}\ \text{J}}{6.626 \times 10^{-34}\ \text{J s}} = 5.22 \times 10^{14}\ \text{s}^{-1}$$

9.27 (a) The wavefunction is $\psi = N\exp(-ax^2/2)$. To normalize it, we have to find N such that

$$1 = \int_{-\infty}^{+\infty} \psi^2(x)\text{d}x = N^2 \int_{-\infty}^{+\infty} e^{-ax^2}\text{d}x = N^2\sqrt{\frac{\pi}{a}}$$

Thus $N^2 (\pi/a)^{1/2} = 1$ or $N = (a/\pi)^{1/4}$

$\psi = (a/\pi)^{1/4}\exp(-ax^2/2)$

(b) Solve equation:

$0 = \text{d}\psi/\text{d}x = -ax(a/\pi)^{1/4}\exp(-ax^2/2)$

The only solution is $x = 0$. Wavefunction reaches maximum for $x=0$, i.e., when oscillator is at its equilibrium position. This is the region where the probability of finding the particle is highest.

9.28 (a) The vibrational frequency of $^{12}C^{16}O$ is:

$$\nu = \frac{1}{2\pi}\left(\frac{k}{\mu}\right)^{1/2}$$

The effective mass of the molecule is:

$$\mu = \frac{(12.011 \times 10^{-3}\ \text{kg mol}^{-1}) \times (15.9994 \times 10^{-3}\ \text{kg mol}^{-1})}{(15.9994 \times 10^{-3}\ \text{kg mol}^{-1} + 12.011 \times 10^{-3}\ \text{kg mol}^{-1}) \times 6.022367 \times 10^{23}\ \text{mol}}$$

$$\mu = 1.13919 \times 10^{-26}\ \text{kg}$$

Therefore, the vibrational frequency of $^{12}C^{16}O$ is:

$$\nu = \frac{1}{2\pi}\left(\frac{k}{\mu}\right)^{1/2}$$

$$\nu = \frac{1}{2\pi}\left(\frac{1860\ \text{kg s}^{-2}}{1.13919 \times 10^{-26}\ \text{kg}}\right)^{1/2}$$

$$\nu = 6.431 \times 10^{13}\ \text{s}$$

(b) And the vibrational wavenumber is:

$$\bar{\nu} = \frac{\nu}{c} = \frac{6.431 \times 10^{13}\ \text{s}}{2.998 \times 10^{10}\ \text{cm s}^{-1}}$$

$$\bar{\nu} = 2145\ \text{cm}^{-1}$$

(c) The vibrational wavenumber of $^{12}C^{16}O$ was calculated in part (b).
Effective mass of $^{13}C^{16}O$, $^{12}C^{18}O$, and $^{13}C^{18}O$ are:

$$\mu_{^{13}C^{16}O} = \frac{\left(13.0033 \times 10^{-3}\ \text{kg mol}^{-1}\right) \times \left(15.9994 \times 10^{-3}\ \text{kg mol}^{-1}\right)}{\left(15.9994 \times 10^{-3}\ \text{kg mol}^{-1} + 13.0033 \times 10^{-3}\ \text{kg mol}^{-1}\right) \times 6.022367 \times 10^{23}\ \text{mol}}$$

$$\mu_{^{13}C^{16}O} = 1.1911 \times 10^{-26}\ \text{kg}$$

$$\mu_{^{12}C^{18}O} = \frac{\left(12.011 \times 10^{-3}\ \text{kg mol}^{-1}\right) \times \left(17.9991 \times 10^{-3}\ \text{kg mol}^{-1}\right)}{\left(17.9991 \times 10^{-3}\ \text{kg mol}^{-1} + 12.011 \times 10^{-3}\ \text{kg mol}^{-1}\right) \times 6.022367 \times 10^{23}\ \text{mol}}$$

$$\mu_{^{12}C^{18}O} = 1.1962 \times 10^{-26}\ \text{kg}$$

$$\mu_{^{13}C^{18}O} = \frac{\left(13.0033 \times 10^{-3}\ \text{kg mol}^{-1}\right) \times \left(17.9991 \times 10^{-3}\ \text{kg mol}^{-1}\right)}{\left(17.9991 \times 10^{-3}\ \text{kg mol}^{-1} + 13.0033 \times 10^{-3}\ \text{kg mol}^{-1}\right) \times 6.022367 \times 10^{23}\ \text{mol}}$$

$$\mu_{^{13}C^{18}O} = 1.25355 \times 10^{-26}\ \text{kg}$$

The corresponding vibrational frequency of $^{13}C^{16}O$, $^{12}C^{18}O$, and $^{13}C^{18}O$ are:

$$\nu_{^{13}C^{16}O} = \frac{1}{2\pi}\left(\frac{1860\ \text{kg s}^{-2}}{1.1911 \times 10^{-26}\ \text{kg}}\right)^{1/2}$$

$$\nu_{^{13}C^{16}O} = 6.289 \times 10^{13}\ \text{s}$$

$$\nu_{^{12}C^{18}O} = \frac{1}{2\pi}\left(\frac{1860\ \text{kg s}^{-2}}{1.1962 \times 10^{-26}\ \text{kg}}\right)^{1/2}$$

$$\nu_{^{12}C^{18}O} = 6.276 \times 10^{13}\ \text{s}$$

$$\nu_{^{13}C^{18}O} = \frac{1}{2\pi}\left(\frac{1860\ \text{kg s}^{-2}}{1.25355 \times 10^{-26}\ \text{kg}}\right)^{1/2}$$

$$\nu_{^{13}C^{18}O} = 6.131 \times 10^{13}\ \text{s}$$

Finally the vibrational wavenumber of $^{13}C^{16}O$, $^{12}C^{18}O$, and $^{13}C^{18}O$ compared to $^{12}C^{16}O$ are shown in the table below:

	$^{13}C^{16}O$	$^{12}C^{18}O$	$^{13}C^{18}O$	$^{12}C^{16}O$
$\bar{\nu}$ /cm^{-1}	2097	2093	2045	2145

9.29 The ionization energy of a hydrogen-like atom is the energy difference for the process of removing the electron from the ground state, n=1 (energy E_1) to a state where the nucleus and electron are infinitely far apart, n=∞ (energy E_∞=0):

$$I = 0 - E_1 = -E_1$$

The energy levels of hydrogen-like systems are

$$E_n = -hcRZ^2/n^2$$

where h, c and R are constants and Z is the atomic number. (Here R is not the gas constant, but the Rydberg constant.)

The ratio of ionization energies of Li^{+2} (Z=3) and He^+ (Z=2) is thus

$$I(Li^{+2})/I(He^+) = E_1(Z{=}3)/E_1(Z{=}2) = 3^2/2^2 = 9/4 = 2.25$$

$$I(Li^{+2}) = 2.25\ I(He^+) = 2.25\ (54.36\ eV) = 122.31\ eV$$

9.30 The value of n is 4 for the N shell. There are16 orbitals: one s, three p, five d and seven f. The general answer is N^2.

9.31 The hydrogen atom ground state wavefunction (1s) is $\psi(r) = N\exp(-r/a_0)$, where N is a constant normalization factor and a_0 is the Bohr radius.

(a) The probability of finding the electron in a small volume at distance r from the nucleus is proportional to

$$\psi^2 = N^2\exp(-2r/a_0)$$

$\psi^2(r)$ is a monotonically decreasing function of r, with maximum value at r=0. To find value of r where ψ^2 drops to 25% of its maximum value, we solve the equation

$$0.25 = \psi^2(r)/\psi^2(0) = (N^2\exp(-2r/a_0))/(N^2) = \exp(-2r/a_0)$$

Taking logarithms gives

$$\ln(0.25) = -2r/a_0$$

$$r = -(a_0/2)\times\ln(0.25) = 0.693\ a_0$$

(b) The radial probability distribution of finding the electron is

$$P(r) = 4\pi r^2\psi^2 = 4\pi r^2 N^2\exp(-2r/a_0) = 4\pi N^2 r^2\exp(-2r/a_0)$$

This function has a maximum at $r = a_0$ (see part c). To find value of r where $P(r)$ drops to 25% of its maximum value we can solve the equation:

$$0.25 = P(r)/P(a_0) = (4\pi N^2 r^2\exp(-2r/a_0))/(4\pi N^2 a_0^2\exp(-2)) = (e^2r^2/a_0^2)\ \exp(-2r/a_0)$$

We can introduce a new variable u = r/a_0 and rewrite this as

$$u^2e^{-2u} = 0.25/e^2 = 3.38\times10^{-2}$$

You can solve this on your calculator. An alternative solution is to plot the left hand side of the above equation and find values of u where the function is equal to 3.38×10^{-2}.

Either way, we find two solutions:

$$u_1 \approx 0.2 \quad \text{and} \quad u_2 \approx 2.6$$

which correspond to

$$r \approx 0.2a_0 \quad \text{and} \quad r \approx 2.6a_0$$

(c) To analyze the behavior of the radial distribution function

$$P(r) = 4\pi N^2 r^2 \exp(-2r/a_0)$$

solve the equation

$$0 = \mathrm{d}P(r)/\mathrm{d}r = 4\pi N^2 \left[(2r \times \exp(-2r/a_0)) - ((2r^2/a_0) \times \exp(-2r/a_0))\right]$$

$$0 = 8\pi N^2 r[1-(r/a_0)] \exp(-2r/a_0)$$

The solutions are $r = 0$ (a minimum) and $r = a_0$ (a maximum). $P(r)$ increases from a value of zero at $r = 0$ to its maximum value at $r = a_0$, followed by a gradual fall of toward zero at large values of r. The most probable distance of an electron from the nucleus in the 1s state corresponds to the maximum of $P(r)$, i.e. $r = a_0$.

9.32 50%, given that it occupies that orbital, but if it is present in that *level*, then the answer is one in six.

9.33 For the hydrogen atom in the 2s state, the wavefunction is

$$\psi(r) = (1/32\pi a_0^3)^{1/2}[2 - (r/a_0)]\exp(-r/2a_0)$$

and the probability of finding the electron in a small volume $\mathrm{d}\tau$ at a distance r from the nucleus is

$$P = \psi^2(r)\,\mathrm{d}\tau = (1/32\pi a_0^3)[2 - (r/a_0)]^2\exp(-r/a_0)\mathrm{d}\tau$$

(a) $\mathrm{d}\tau = 1\ \mathrm{pm}^3$

i) $P = \psi^2(0)\,\mathrm{d}\tau = (1/32\pi a_0^3)[2]^2(1)\mathrm{d}\tau = (4\ \mathrm{pm}^3)/(32\pi(52.9177\ \mathrm{pm})^3) = 2.7\times10^{-7}$

ii) $\mathrm{P} = \psi^2(a_0)\,\mathrm{d}\tau = (1/32\pi a_0^3)[1]^2(\mathrm{e}^{-1})\mathrm{d}\tau = (\mathrm{e}^{-1}\ \mathrm{pm}^3)/(32\pi(52.9177\ \mathrm{pm})^3)$
$=2.5\times10^{-8}$

iii) $P = \psi^2(2a_0)\,\mathrm{d}\tau = (1/32\pi a_0^3)[0]^2(\mathrm{e}^{-2})\mathrm{d}\tau = 0$

(b) The radial distribution function

$\mathrm{P(r)} = 4\pi r^2\ \psi^2(r) = (r^2/8a_0^3)[2 - (r/a_0)]^2\exp(^-r/a_0)$

$= (1/8a_0)\ (r/a_0)^2[2 - (r/a_0)]^2\exp(-r/a_0) = (8/a_0)f(u)$

where

$$f(u) = u^2(2-u)^2\mathrm{e}^{-u} \text{ with } u = r/a_0$$

Plotting $f(u)$ shows that the function has two minima, at $u = 0$ and $u = 2$ and two maxima, at $u \approx 0.8$ and $u \approx 5.2$.

The corresponding values of r are thus: minima at $r = 0$ and $r = 2a_0$,

maxima at $u \approx 0.8a_0$ and $u \approx 5.2a_0$

(c) Calculate the derivative

$$dP(r)/dr = (2r/8a_0^3)[2-(r/a_0)]^2\exp(-r/a_0)$$
$$+ (r^2/8a_0^3)(-2/a_0)[2-(r/a_0)]\exp(-r/a_0)$$
$$+ (r^2/8a_0^3)[2-(r/a_0)]^2(-1/a_0)\exp(-r/a_0)$$
$$= (r/8a_0^3)[2-(r/a_0)]\{2[2-(r/a_0)]-(2r/a_0)-(r/a_0)\,[2-(r/a_0)]\}\,(\exp(-r/a_0)$$
$$= (r/8a_0^3)[2-(r/a_0)][(r/a_0)^2-(6r/a_0)+4]\,(\exp(-r/a_0)$$

The equation d$P(r)$/dr = 0 has four solutions. Two of them, $r = 0$ and $r = 2a_0$ are easy to find by inspection; they correspond to minima of the distribution – check by plotting.
The two remaining solutions are obtained by solving the quadratic equation in the square bracket, and correspond to the two maxima at
$r = (3 - 5^{1/2})a_0 \approx 0.76a_0$ and $r = (3 + 5^{1/2})a_0 \approx 5.24a_0$ as previously determined in (b) from plot.
The value of the Bohr radius a_0 = 52.9177 pm was used.

9.34 (a) The radial function corresponding to the 3s orbital of an H atom is proportional to $6 - 2\rho + 1/9\,\rho^2$ where $\rho = 2\,r/a_0$.
To locate the radial nodes we will set $6 - 2\rho + 1/9\,\rho^2 = 0$, solve for ρ, and then evaluate r using a_0 = 52.92 pm.
The roots of the quadratic equation are: ρ= 14.196 and ρ= 3.80
By using $\rho = 2\,r/a_0$, the radial nodes in the 3s orbital of an H atom are located at 101 pm and 376 pm.
(b) The radial function corresponding to the 4s orbital of an H atom is proportional to $24 - 36\rho + 12\,\rho^2 - \rho^3$.
The roots of the cubic equation are: ρ= 3.741, ρ= 13.189 and ρ= 31.03.
By using $\rho = 2\,r/a_0$, the radial nodes in the 4s orbital of an H atom are located at 99 pm, 349 pm and 821 pm.

9.35 The nodal planes are where the angular term is zero. In our case we look for solutions to $\sin(\theta)\cos(\theta) = 0$, which happens when either $\sin(\theta) = 0$ or $\cos(\theta) = 0$. The range θ of is from 0 to π, so the possible solutions are

θ= 0, π/2 and π

θ= π/2 corresponds to the xy plane, so this is a nodal plane of this orbital.

The angles θ= 0 and π correspond to the z axis, so all of the infinite number of planes that pass through that axis satisfy our condition. We can narrow things down by writing down the full angular form of these kinds of orbitals:

$3d_{xz}$: $\sin(\theta)\cos(\theta)\cos(\varphi)$ - yz is the second nodal plane
$3d_{yz}$: $\sin(\theta)\cos(\theta)\sin(\varphi)$ - xz is the second nodal plane

9.36 The orbital angular momentum of an electron in different orbitals can be evaluated by using the equation $J = \{l(l+1)\}^{1/2}\hbar$.

The table shows the value of J of different orbitals:

	l	J
1s	0	0
3s	0	0
3d	2	$6^{1/2}\hbar$
2p	1	$2^{1/2}\hbar$
3p	1	$2^{1/2}\hbar$

The number of radial nodes depends on both n and l, with the formula, # radial nodes = n – l – 1 and the number of angular nodes is equal to l.

The numbers of angular and radial nodes in each orbital are listed in the table below:

	1s	3s	3d	2p	3p
angular	0	0	2	1	1
radial	0	2	0	0	1

9.37 There are $2l+1$ states in the subshell, corresponding to $m = -l,\ldots,0,\ldots,l$, each can be occupied by two electrons with opposite spins, $2(2l+1)$ electrons altogether.

(a) $2(2l+1) = 2(2\times0 + 1) = 2$

(b) $2(2l+1) = 2(2\times3 + 1) = 14$

(c) $2(2l+1) = 2(2\times5 + 1) = 22$

9.38 (a)

H	He																										
Li	Be																			B	C	N	O	F	Ne	Na	Mg
Al	Si																			P	S	Cl	Ar	K	Ca	Sc	Ti
V	Cr																										

(b) Mg and Ti would be noble gases.

(c) Probably N because it would have 5 electrons in the valence shell – which would be half full, like C.

9.39 The occupied orbitals are:

Fe [Ar]$3d^6 4s^2$

Fe^{2+} [Ar]$3d^6$

Fe^{3+} [Ar]$3d^5$

We expect Fe^{3+} to be smaller than Fe^{2+}. The physical reason is the strong tendency of electrons within each subshell to repel each other. Thus, the presence of the extra electron in Fe^{2+} will lead to extra repulsion, leading to increase of average electron distance from the nucleus and increased size of the ion.

9.40 Ionization energies were obtained from the American Chemical Society Web site.

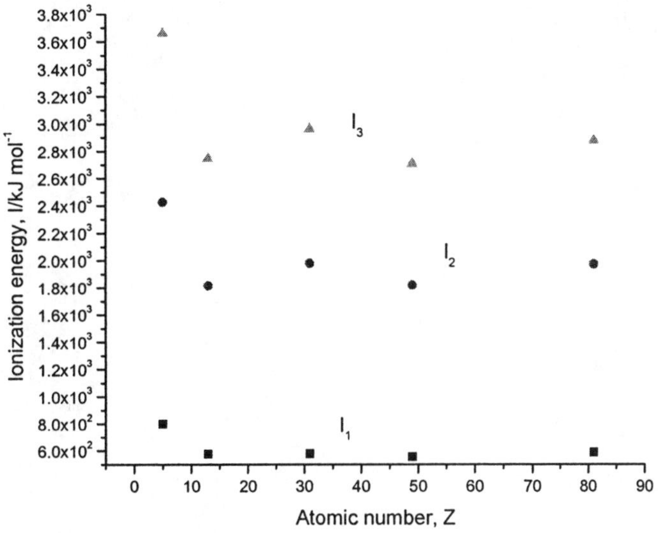

Elements from Al to Thallium, with the exception of Gallium, follow a similar pattern in the sense that the magnitudes of the first, second and third ionization energies are very similar. As we would expect, the energy needed to remove electrons increase as the number of electrons to be removed increase. There is not much of a difference in the values of I_3 for Al and Thallium to justify a preference for an oxidation number. However in the Thallium, the shielding effect by *d* and *f* electrons might play an important role in determining the oxidation state of Tl.

9.41 The ionization energy I_{A^-} of the anion A^- is the energy change for the process:

$$A^-(g) \rightarrow A(g) + e^-(g)$$

$$I_{A^-} = E(A,g) + E(e^-,g) - E(A^-,g)$$

The electron affinity E_A of the parent atom A is the energy change for the process

$$A(g) + e^-(g) \rightarrow A^-(g)$$

$$E_A = E(A^-,g) - E(A,g) - E(e^-,g)$$

Clearly, this second process is the reverse of the first (and *vice versa*), so that

$$I_{A^-} = -E_A$$

9.42 The increasing magnitude of the equilibrium constant is parallel to the decreasing size of cations. The nature of the interaction between ions is columbic and its magnitude increases with decreasing distance between ions.

Chapter 10: The Chemical Bond

10.8 The wavefunctions for each bond are:

$$\Psi_{\pi} = \Psi_{p_{x,y}}^{N_A}(1)\Psi_{p_{x,y}}^{N_B}(2) + \Psi_{p_{x,y}}^{N_A}(2)\Psi_{p_{x,y}}^{N_B}(1)$$ (i.e., one π from p_x's, one from p_y's)

$$\Psi_{\sigma} = \Psi_{p_z}^{N_A}(1)\Psi_{p_z}^{N_B}(2) + \Psi_{p_z}^{N_A}(2)\Psi_{p_z}^{N_B}(1)$$

10.9

$$E = F/r = kq_1q_2/r = \frac{\left(1.602\times10^{-19}\,\mathrm{C}\right)^2}{4\pi(8.854\times10^{-12}\,\mathrm{C^2/J\,m})(74.1\times10^{-12}\,\mathrm{m})}(6.022\times10^{23}\,\mathrm{mol^{-1}})$$

$$= 1.88\times10^{6}\,\mathrm{J/mol}$$

10.10 The VB description of SO_2 is: two σ bonds between Ssp^2 and Osp^2 hybrid orbitals, and a π bond between unhybridized $S3p_z$ and $O2p_z$ orbitals.
The VB description of SO_3 is: three σ bonds between Ssp^2 and Osp^2 hybrid orbitals, and a π bond between unhybridized $S3p_z$ and $O2p_z$ orbitals.

10.11 O(*sp*^{2}), N (*sp*^{2}), next 2O's (*sp*^{3})

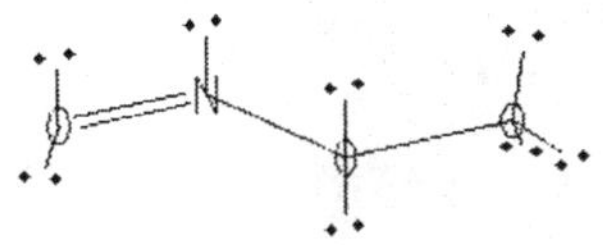

10.12 The structure of 11-*cis* retinal drawn below is showing carbon atoms (1-11) that use sp^2 hybrid atomic orbitals to form σ bonds with carbon and hydrogen atoms. There are six π bonds, one of which is between carbon numbered 11 and the oxygen atom (which is also sp^2 hybridized). The other carbon atoms are sp^3.

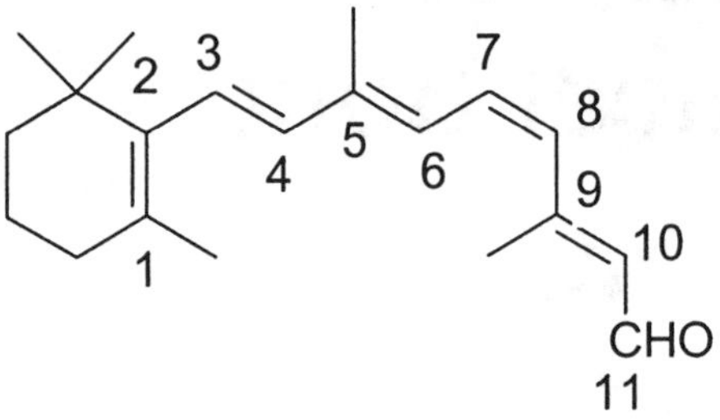

10.13

$$\int_V h_1 h_2 \mathrm{d}\tau = \int_V (s + p_x + p_y + p_z)(s - p_x - p_y + p_z)\mathrm{d}\tau = 1 - 1 - 1 + 1 = 0$$

All cross terms = 0 due to orthogonality of the atomic orbitals.

10.14 To normalize the sp^2 hybrid orbital we need to evaluate the integral:

$$\int \psi^2 \mathrm{d}\tau = 1 = \frac{1}{3}\int \left(s + \sqrt{2}p\right)^2$$

$$\frac{1}{3}\int \left(s^2 + 2sp\sqrt{2} + 2p^2\right)\mathrm{d}\tau = 1$$

$$\frac{1}{3}\left(1 + 0 + 2\right) = 1$$

The integrals $\int s^2 \mathrm{d}\tau$ and $\int p^2 \mathrm{d}\tau$ are equal to 1 and $\int 2sp\sqrt{2}\mathrm{d}\tau$ is zero due to the orthogonally of the atomic orbitals.

10.15 $h_3 = s - p_x + p_y - p_z$

10.16

$$\int \psi^2 \mathrm{d}\tau = 1 = N^2 \int \left(\psi_{\text{cov}} + \lambda\psi_{\text{ion}}\right)^2 \mathrm{d}\tau$$

$$N^2 \int \left(\psi^2_{\text{cov}} + 2\lambda\psi_{\text{cov}}\psi_{\text{ion}} + \lambda^2\psi^2_{\text{ion}}\right)\mathrm{d}\tau$$

$$N^2\left(1 + 2\lambda S + \lambda^2\right) = 1$$

The term $\int \psi_{\text{cov}}\psi_{\text{ion}}\mathrm{d}\tau$ is the overlap integral S.

The normalization constant is then:

$$N = \left[\frac{1}{\left(1 + 2\lambda S + \lambda^2\right)}\right]^{1/2}$$

10.17 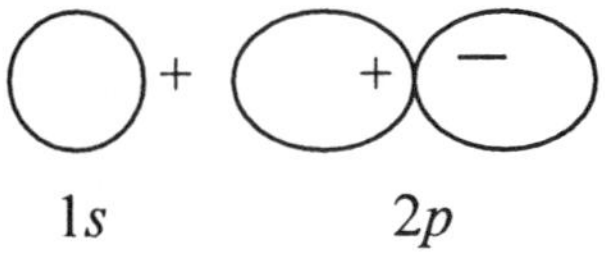

S is a maximum at $R = 2.104\ a_0$.

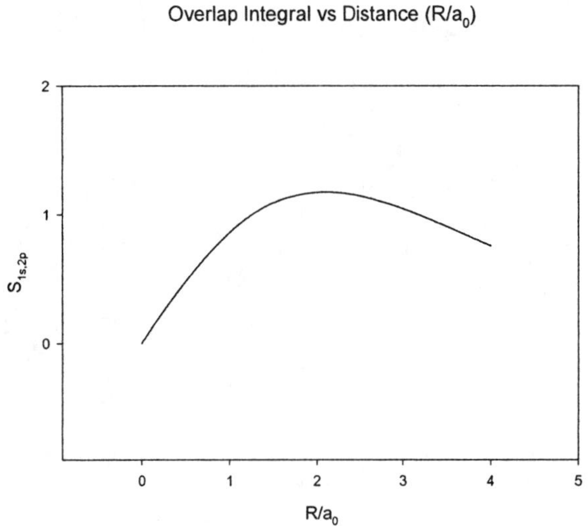

10.18 To find a linear combination of the orbitals A and B that is orthogonal to the given combination, we need to evaluate the integral $\int \psi_a \psi_b d\tau = 0$.

We will designate $\psi_a = aA + bB$ and $\psi_b = N(0.145A + 0.844B)$

Then,

$$\int (aA + bB)\ N(0.145A + 0.844B) d\tau = 0$$

$$\int \left(aNA^2 0.145 + aABN0.844 + bABN0.145 + bNB^2 0.844\right) d\tau = 0$$

The integrals $\int A^2 d\tau$ and $\int B^2 d\tau$ are equal to 1 and $\int AB d\tau$ is the overlap integral S, then after collecting terms:

$$a(0.145 + S0.844) = -\ b(S0.145 + 0.844)$$

Or,

$$\frac{a}{b} = -\frac{(S0.145 + 0.844)}{(0.145 + S0.844)}$$

Assuming the functions A and B are orthogonal (i.e. S=0), then:

$\frac{a}{b} = -\frac{.844}{.145}$. When a= −0.844 and b= 0.145, we get:

N(−0.844A+0.145B), which is indeed orthogonal.

10.19 Both electrons on one atom means ionic character. Therefore, the chance is $(.150)^2 = 0.0225$ to 1, or 22.5 out of 1000.

10.20 (a) A graph of E versus a is shown below. The minimum energy is obtained at a=1.25, corresponding to an energy of -2.2899×10^{-38}J.

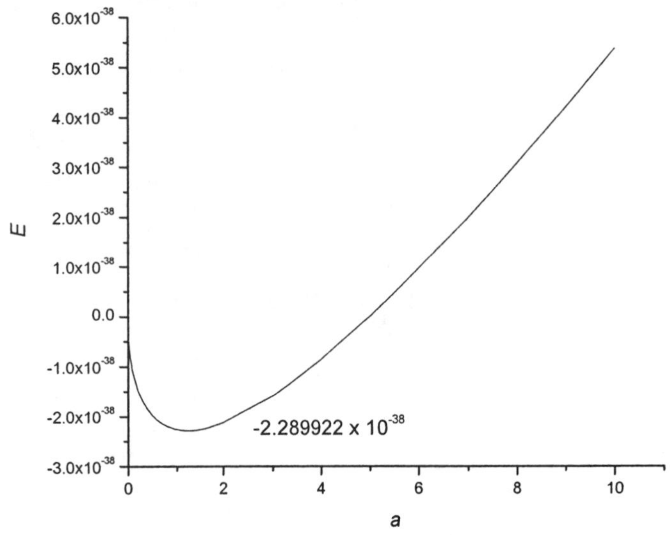

(b) If we determine the minimum energy by differentiation, we need to set the derivative $\frac{dE}{da}$ to zero and solve for the value of a as shown below:

$$\frac{dE}{da} = \frac{3\hbar^2}{2\mu} - e^2\sqrt{\frac{2}{a\pi}} = 0$$

Solving for a:

$$\frac{3\hbar^2}{2\mu} = e^2\sqrt{\frac{2}{a\pi}}$$

$$\frac{2\mu}{3\hbar^2} = \frac{\sqrt{a\pi}}{e^2\sqrt{2}}$$

$$a = \frac{1}{\pi}\left(\frac{2\mu\sqrt{2}e^2}{3\hbar^2}\right)^2 = \frac{8\mu^2 e^4}{9\pi\hbar^4}$$

Substituting of the corresponding parameters into the latest equation, we get a=1.249. A value of 9.10429×10^{-31} kg was used for the effective mass of the H atom.

10.21 There are no other resonance structures for benzene. The MO picture is one of a low energy bonding ground state and a higher energy, doubly degenerate bonding state. Above this is another doubly degenerate state but it is antibonding and the highest energy state is nondegenerate and antibonding with the maximum number of nodes. See Figure 10.36 of the text. In simple descriptions, the Kekule resonance structures show how delocalization stabilizes the ring.

10.22 (a) Addition of the squares of the coefficients should equal 1, which is the case if the coefficients are cosθ and sinθ.

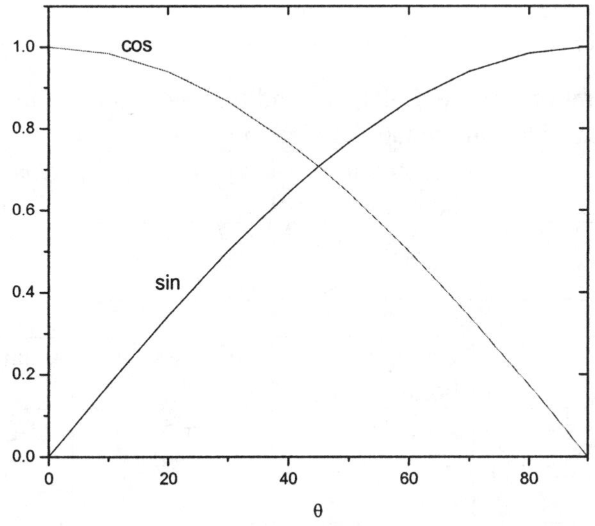

(b) To verify that ψ is normalized to 1, we evaluate the following integral:

$$\int \psi^2 \mathrm{d}\tau = 1 = \int (\psi_A \cos\theta + \psi_B \sin\theta)^2 \mathrm{d}\tau$$

$$\int (\psi_A^2 \cos^2\theta + \psi_B^2 \sin^2\theta + 2\psi_A\psi_B \cos\theta \sin\theta)\mathrm{d}\tau = 1$$

Since the integrals $\int \psi_A^2 \mathrm{d}\tau$ and $\int \psi_B^2 \mathrm{d}\tau$ are equal to 1 and $\int \psi_A \psi_B \mathrm{d}\tau$ is zero, then the expression above reduces to:

$$\int (\cos^2\theta + \sin^2\theta)\mathrm{d}\tau = 1$$

Since $\cos^2\theta + \sin^2\theta = 1$ we have demonstrated ψ is normalized to 1.

(c) In a homonuclear diatomic molecule, the squares of the coefficients must be the same, so cosθ is equal to sinθ at 45° as shown in the graph in part (a).

10.23 This is depicted in Figure 10.31 of the text, where d_{xz} and d_{yz} orbitals may combine with p_y and p_x orbitals to form π orbitals.

10.24 (a) H_2^-: $1\sigma^2 1\sigma^{*1}$
(b) N_2: $1\sigma^2 2\sigma^{*2}\, 1\pi^4 3\sigma^2$
(c) O_2: $1\sigma^2 2\sigma^{*2}\, 3\sigma^2 1\pi^4 2\pi^{*2}$

10.25 (a) CO: $1\sigma^2 2\sigma^{*2}\, 1\pi^4 3\sigma^2$
(b) NO: $1\sigma^2 2\sigma^{*2}\, 1\pi^4 3\sigma^2 2\pi^*$
(c)CN^-: $1\sigma^2 2\sigma^{*2}\, 1\pi^4 3\sigma^2$

10.26 The electronic configurations of the molecules and their corresponding ions after addition or removal of an electron are tabulated below:

N_2	N_2^-	Change in bond order	N_2^+	Change in bond order
$1\sigma^2 2\sigma^{*2}$ $1\pi^4 3\sigma^2$	$1\sigma^2 2\sigma^{*2}$ $1\pi^4 3\sigma^2 2\pi^*$	3 –2.5 = 0.5	$1\sigma^2 2\sigma^{*2}$ $1\pi^4 3\sigma$	3 –2.5 = 0.5

O_2	O_2^-	Change in bond order	O_2^+	Change in bond order
$1\sigma^2 2\sigma^{*2}$ $3\sigma^2 1\pi^4 2\pi^{*2}$	$1\sigma^2 2\sigma^{*2}$ $3\sigma^2 1\pi^4 2\pi^{*3}$	2 –1.5 = 0.5	$1\sigma^2 2\sigma^{*2}$ $3\sigma^2 1\pi^4 2\pi^*$	2 –2.5 = –0.5

NO	NO^-	Change in bond order	NO^+	Change in bond order
$1\sigma^2 2\sigma^{*2}$ $1\pi^4 3\sigma^2 2\pi^*$	$1\sigma^2 2\sigma^{*2}$ $1\pi^4 3\sigma^2 2\pi^{*2}$	2.5 –2.0 = 0.5	$1\sigma^2 2\sigma^{*2}$ $1\pi^4 3\sigma^2$	2.5 –3 = –0.5

C_2	C_2^-	Change in bond order	C_2^+	Change in bond order
$1\sigma^2 2\sigma^{*2}\, 1\pi^4$	$1\sigma^2 2\sigma^{*2}$ $1\pi^4 3\sigma^1$	2 –2.5 = –0.5	$1\sigma^2 2\sigma^{*2}\, 1\pi^3$	2 –1.5 = 0.5

F_2	F_2^-	Change in bond order	F_2^+	Change in bond order
$1\sigma^2 2\sigma^{*2}$ $3\sigma^2 1\pi^4 2\pi^{*4}$	$1\sigma^2 2\sigma^{*2}$ $3\sigma^2 1\pi^4 2\pi^{*4} 4\sigma^*$	1 –0.5 = 0.5	$1\sigma^2 2\sigma^{*2}$ $3\sigma^2 1\pi^4 2\pi^{*3}$	1 –1.5 = –0.5

CN	CN^-	Change in bond order	CN^+	Change in bond order
$1\sigma^2 2\sigma^{*2}$ $1\pi^4 3\sigma^1$	$1\sigma^2 2\sigma^{*2}$ $1\pi^4 3\sigma^2$	2.5 −3.0 = −0.5	$1\sigma^2 2\sigma^{*2} 1\pi^4$	2.5 −1.5 = 0.5

The molecules O_2, NO and F_2 will be stabilized by cation formation. The molecules CN and C_2 will be stabilized by anion formation.

10.27 The parities alternate from the ground state up starting with g. Thus, g,u,g,u.

10.28 (a) g, u, g, u
(b) Odd quantum number v corresponds to odd parity, and even quantum number v corresponds to even parity.

10.29 1π: u ; 2 × 2π: g; 2 × 3π: u; 4π: g.

10.30 In Exercise 10.26, we wrote the electron configurations of NO and N_2 and determined their corresponding bond orders are 2.5 and 3 respectively. Nitrogen is likely to have the greater bond dissociation energy and the shorter bond length.

10.31 The bond orders are as follows: $O_2^{2-} = 1.$ $O_2^- = 1.5.$ $O_2 = 2.$ $O_2^+ = 2.5.$
Therefore, order of the bond lengths is: $O_2^{+2} > O_2 > O_2^- > O_2^{-2}$

10.32 (a) The molecular orbital energy level diagram of ethene is:

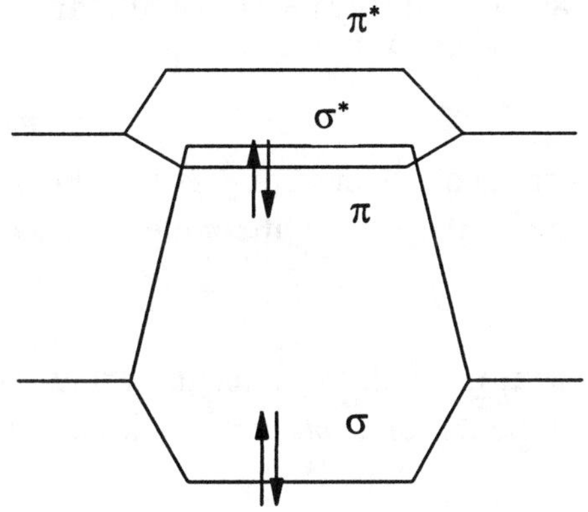

(b) The molecular orbital energy level diagram of ethyne is:

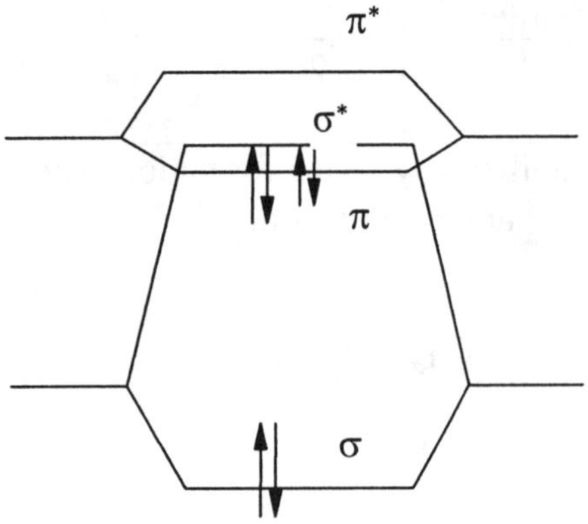

10.33 

ψ_1 ψ_2

$$E_{\text{min}} = \frac{(3^2-2^2)h^2}{8ma^2} = \frac{5(6.626\times10^{-34}\,\text{J s})^2}{8\times(9.11\times10^{-31}\text{kg})\times(4\times140\times10^{-12}\text{m})^2} = 9.60\times10^{-19}\,\text{J}$$

(b) There are no sketches: HOMO has 3 nodes, LUMO has 4.

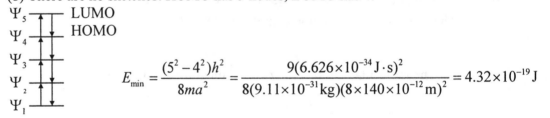

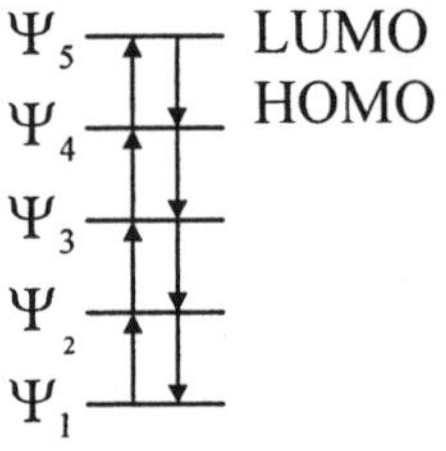

$$E_{\text{min}} = \frac{(5^2-4^2)h^2}{8ma^2} = \frac{9(6.626\times10^{-34}\,\text{J}\cdot\text{s})^2}{8(9.11\times10^{-31}\text{kg})(8\times140\times10^{-12}\text{m})^2} = 4.32\times10^{-19}\,\text{J}$$

10.34 The energy of the binding π molecular orbital of ethane is $\alpha + \beta$. Since there are two electrons to accommodate in the π molecular orbital, the π-electron binding energy, E_π is:

$$E_\pi = 2(\alpha + \beta) = 2\alpha + 2\beta$$

In the case of butadiene, there are two electrons in the π molecular orbital with an energy of $\alpha + 1.62\beta$ and two electrons in the π molecular orbital with an energy of $\alpha + 0.62\beta$.

Thus, the *π-electron binding energy*, E_π is:

$$E_\pi = 2(\alpha + 1.62\beta) + 2(\alpha + 0.62\beta) = 4\alpha + 4.48\beta$$

It is lower.

10.35 (a)The carbon atoms are sp^2 hybridized and the $p\pi$ orbitals extend above and below the plane of the ring. Bonding takes place by overlap of the p orbitals.

(b)

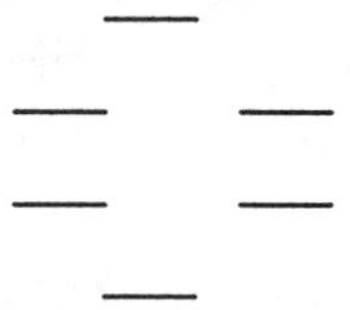

$\alpha{-}2\beta$

$\alpha{-}\beta$

$\alpha{+}\beta$

$\alpha{+}2\beta$

(c) $E_{deloc} = 6\alpha + 8\beta - 3(2\alpha + 2\beta) = 2\beta$

(d) anion: configuration $= 1\pi^2 2\pi^2 3\pi^2 4\pi^1$ $E_{deloc} = 7\alpha + 7\beta - 3(2\alpha + 2\beta) =$ $\alpha{+}\beta$

cation: configuration $= 1\pi^2 2\pi^2 3\pi^1$ $E_{deloc} = 5\alpha + 7\beta - 3(2\alpha + 2\beta) = -\alpha{+}\beta$

10.36 The energy level diagram for $[Fe(CN)_6]^{3-}$ is:

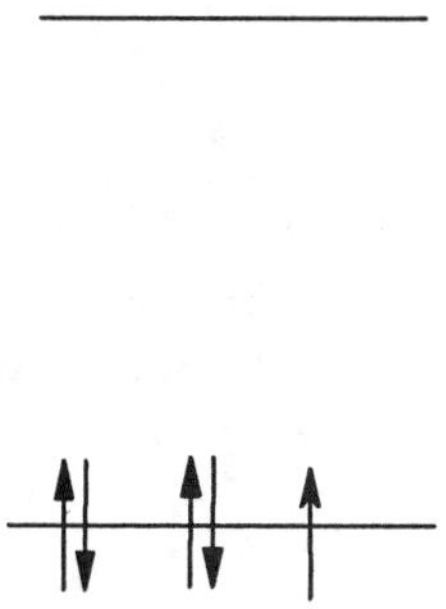

And the number of unpaired electrons in the complex is 1.

The energy level diagram for $[Fe(H_2O)_6]^{3+}$ is:

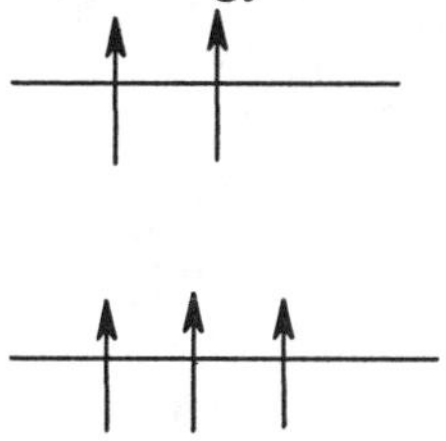

And the number of unpaired electrons in the complex is 5.

10.37 When the number of electrons = 4, 5, 6 or 7.

10.38 A square planar geometry would be the most probable for Ni^{+2} ion since it has to accommodate eight electrons.

10.39 (a) Ligands such as Cl^- supply electrons and are Lewis π bases (*π-donor* orbitals). In the case of Cl^-, the ligand π orbitals are full and lie below the metal orbitals. The effect of the bonding is then to decrease the ligand field splitting (Δ). On the other hand, Lewis π acids such as CO with initially empty π antibonding orbitals (*π-acceptors*) lie above the *d* electron energies. The effect of π overlap is then to increase Δ.

(b) O_2 has the electron configuration $1\sigma^2 1\sigma^{*2} 2\sigma^2 1\pi^4 1\pi^{*2}$ and has a half-filled π-antibonding orbital and is thus a Lewis π acid. As such, it is expected to be a somewhat moderate-field ligand.

(c) Based upon the case studies mentioned, it is seen that O_2 bonds reversibly to the Fe(II) atom of hemoglobin, which means that the bonding is only moderately strong. On the other hand, CO is a strong field ligand and therefore binds very strongly to the Fe(II) atom. This bonding is for all practical purposes irreversible, and CO forms a very stable complex with hemoglobin which does not allow for the transport of O_2.

10.40 As discussed on page 430 in the textbook, if the construction of LCAO-MOs is carried out by replacing exponential functions by a sum of Gaussian functions, the product of two Gaussians is another Gaussian function that is centered between the initial functions.

Chapter 11: Macromolecules and Self-Assembly

11.13 We follow the procedure from Illustration 11.1 in text. Start with Equation 11.3 to obtain molecular mass M from a sedimentation equilibrium:

$$M = (\text{slope of } \ln c \text{ vs. } r^2 \text{ plot}) \times (2RT)/b\omega^2$$

where R is the gas constant, T the temperature, ω the angular velocity and b the buoyancy factor. In our case:

$$\text{slope} = 729 \text{ cm}^{-2} = 7.29\times10^6 \text{ m}^{-2}$$

For each full rotation, there is an angle change of 2π, so a rotation rate of 50,000 rpm gives angular velocity:

$$\omega = 2\pi(50{,}000)/(60 \text{ s}) = 5.236\times10^3 \text{ s}^{-1}$$

(Radians are dimensionless, so radian/s is the same as 1/s).

The buoyancy factor

$$b = 1 - \rho v_s = 1 - (1.00 \text{ g cm}^{-3})(0.61 \text{ cm}^3 \text{ g}^{-1}) = 0.39$$

$$M = (7.29\times10^6 \text{ m}^{-2})(2)(8.314 \text{ J mol}^{-1} \text{K}^{-1})(300 \text{ K})/(5.236\times10^3 \text{ s}^{-1})^2(0.39)$$

$$= 3.40 \times 10^3 \text{ kg mol}^{-1}$$

$$1 \text{ J} = 1 \text{ kg m}^2\text{s}^{-2}$$

11.14 In Derivation 11.1, the drift speed of a particle, s, is found by equating two forces: fs and $m_{\text{eff}}r\omega^2$. Given that in this exercise the particle is under the influence of gravity alone, $m_{\text{eff}}r\omega^2$ has to be replaced by $m_{\text{eff}}g$. The frictional coefficient, f, is found by combining the following equations:

$$f = \frac{k\text{T}}{D} \text{ defined in Derivation 11.1 and } D = \frac{k\text{T}}{6\pi\eta a} \text{ (Equation 8.6)}$$

The effective mass, m_{eff}, is defined in Derivation 11.1 as equal to bm.

After combining all equations mentioned above we have:

$$m_{\mathit{eff}}\, g = fs$$

$$bmg = 6\pi\eta as$$

The buoyancy correction, b, is equal to $1 - \rho v_s$. Introducing this definition we get:

$$(1 - \rho v_s)mg = 6\pi\eta as$$

After distributing *mg*:

$$mg - m\rho v_s g = 6\pi\eta as$$

Where *m*, the mass of the particle, is equal to the volume of the particle *v*, times its density ρ_p. The specific volume of the particle v_s, multiplied by *m*, is equal to the volume of the particle.

Thus,

$$v\rho_p g - \rho v g = 6\pi\eta as$$

After collecting terms:

$$vg\left(\rho_p - \rho\right) = 6\pi\eta as$$

Assuming the particle is spherical,

$$\frac{4}{3}\pi a^3 g\left(\rho_p - \rho\right) = 6\pi\eta as$$

Finally, solve for *s*.

$$s = \frac{2a^2 g}{9\eta}\left(\rho_p - \rho\right) =$$

$$s = \frac{2 \times \left(20 \times 10^{-6}\ \text{m}\right)^2 \times 9.81\ \text{m s}^{-2}}{9 \times 8.9 \times 10^{-4}\ \text{kg m}^{-1}\ \text{s}^{-1}}\left(1750 - 1000\right)\text{kg m}^{-3}$$

$$s = 7.3 \times 10^{-4}\ \ \text{m s}^{-1}$$

11.15 The relation between molar mass *M*, sedimentation constant *S* and diffusion constant *D* is given by Equation 11.1:

$M = SRT/bD$

where *R* is the gas constant, *T* the temperature, and *b* the buoyancy factor.

$b = 1 - \rho v_s = 1 - (1.06\ \text{g cm}^{-3})(\ 0.656\ \text{cm}^3\ \text{g}^{-1}) = 0.305$

$M = (3.2\times10^{-13}\ \text{s})(8.314\ \text{J mol}^{-1}\text{K}^{-1})(293\ \text{K})/\ (8.3\times10^{-11}\ \text{m}^2\ \text{s}^{-1})(0.305)$

$= 31$ kg/mol

where 1 Sv = 3.2×10^{-13} s.

11.16 Rearranging Equation 11.2, we can solve for the angular frequency as shown below:

$$\omega = \sqrt{\frac{2RT}{\left(r_2^2 - r_1^1\right)Mb}\ln\frac{c_2}{c_1}}$$

$$\omega = \sqrt{\frac{2 \times 8.3145\ \text{J K}^{-1}\ \text{mol}^{-1} \times 298\ \text{K}}{\left[(0.070)^2 - (0.050)^2\right]\text{m}^2 \times 10^2\ \text{kg mol}^{-1} \times 0.25}\ln 5}$$

$$\omega = 3.6 \times 10^2\ \text{s}^{-1}$$

Using the definition of angular frequency as $2\pi v$, we solve for *v* and obtain the speed of operation as 58 Hz or 3.4×10^2 rpm.

11.17 (a) Plotting the number of base pairs n_{bp} vs. t^2 gives a straight line. To explain this, we assume that each kind of DNA molecule forms an ion with the same overall charge; e.g. that all DNAs would have $z = 1$. Further, we can assume that the mass of a DNA fragment is propotional to the number of base pairs. Thus,

$m \propto n_{bp}$ for DNA and

$m/z \propto n_{bp}$ for fixed value of z for all fragments.

In a MALDI-TOF experiment, we know that the relation between m/z and time of flight is

$m/z \propto t^2$

Thus, for our DNA fragments

$n_{bp} \propto m/z \propto t^2$

and we will expect the plot of n_{bp} vs. t^2 to be a straight line.

(b) Fitting the provided data to a straight line

$t^2 = a + b \times n_{bp}$ gives $a = 482.30$ $b = 115.61$

Thus, for $n_{bp} = 238$, we expect $t^2 = 482.30 + (115.61)(238) = 27{,}997.5$

$t = 167.3\ \mu s$

The units are same as in data.

11.18 A two-dimensional lattice in terms of 10, 01, 11, 12, 23, 41 and 4-1 is shown below:

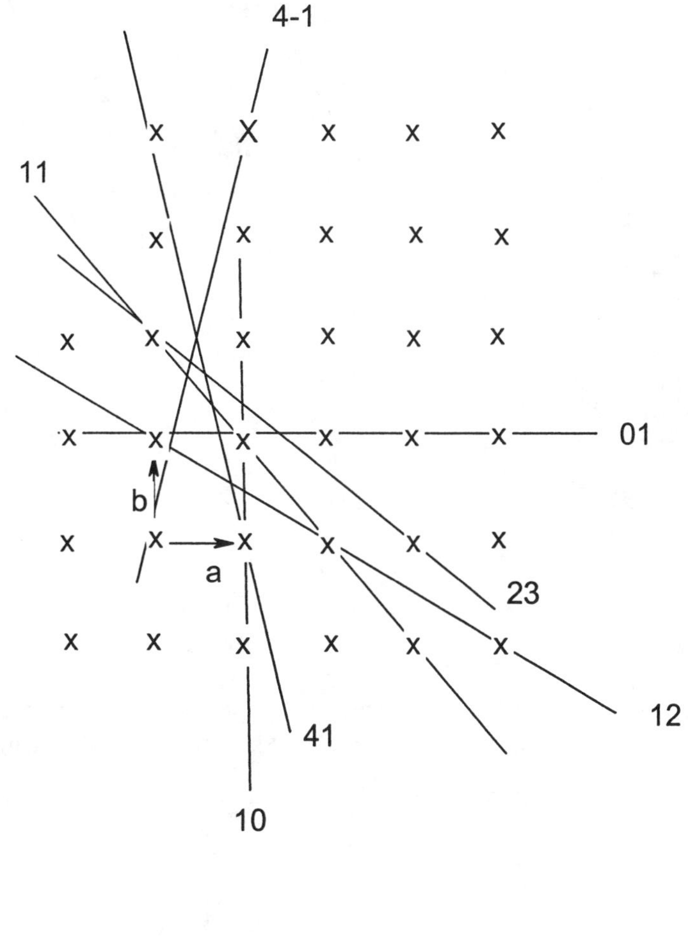

11.19 The two-dimensional grid with axes *a* and *b* forming a 60° angle is shown below. Taking *a* as the horizontal axis and *b* as the vector forming the 60° angle with *a*:

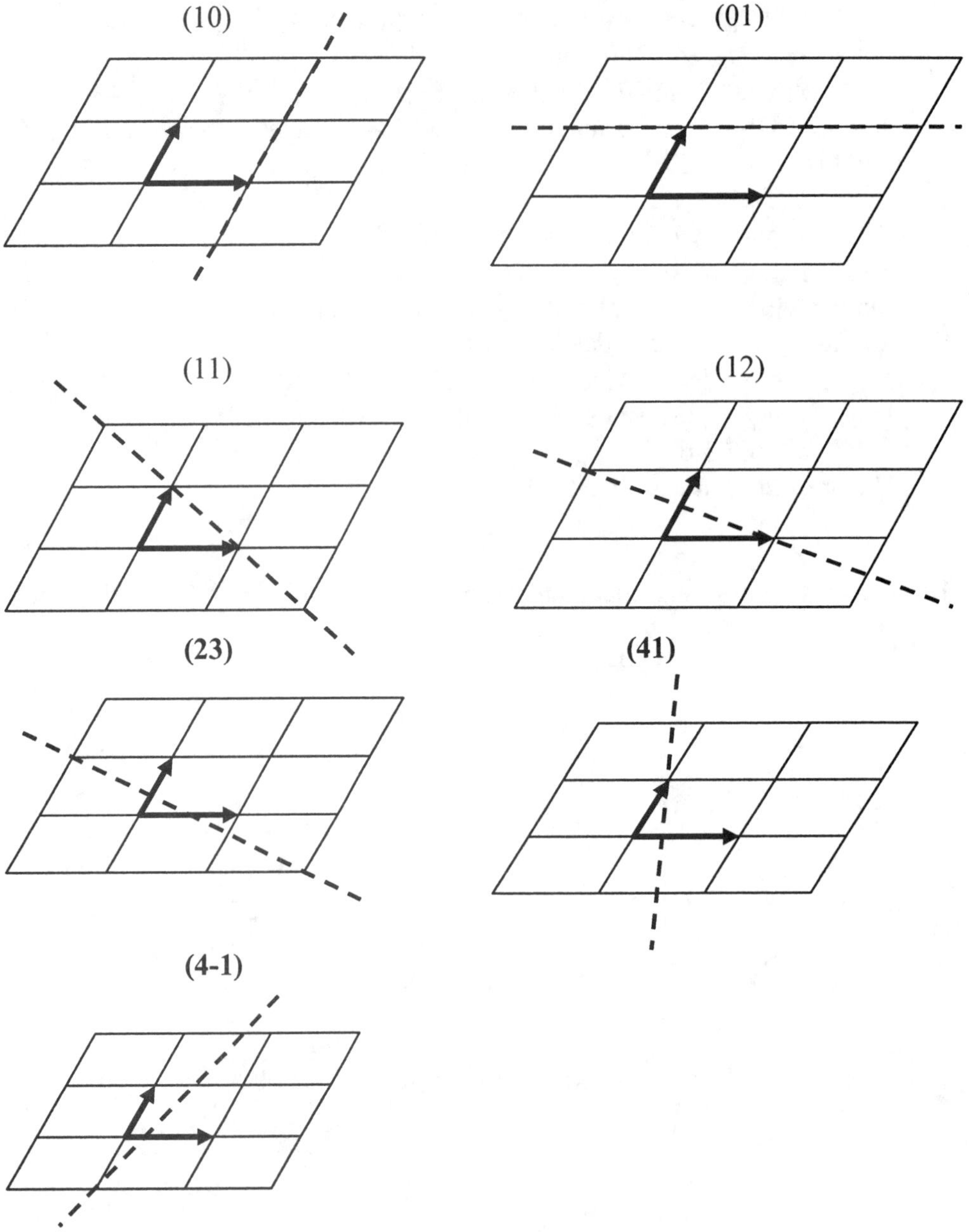

11.20 As discussed in Section 11.3, the Miller indices of the planes are the reciprocals of the given coordinates with fractions cleared. The corresponding Miller indices are then: (326), (111), (122), and $(3\bar{2}\bar{2})$.

11.21 In the orthorhombic unit cell all angles are 90°, and all axes are of different length. Let yz be the plane of the paper, with the y-axis horizontal and z vertical, and x point toward us perpendicular to the plane. The positions of several planes are drawn below.

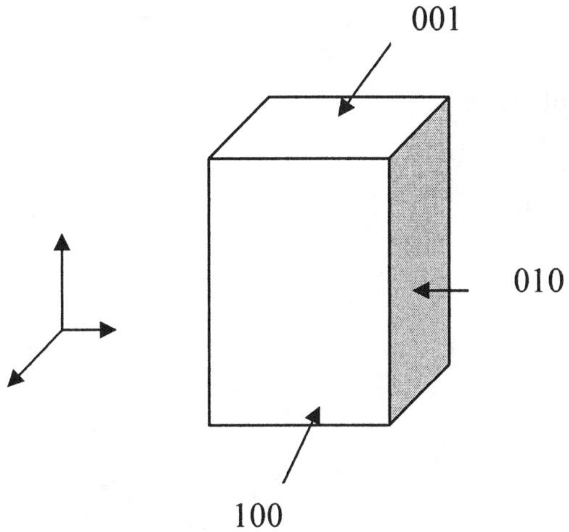

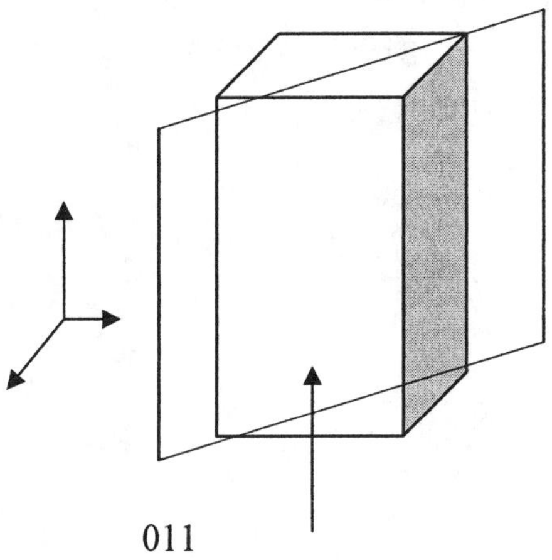

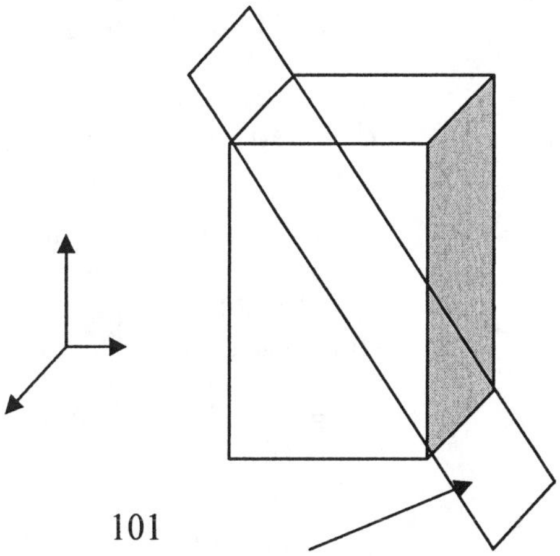

11.22 The distance between planes is calculated using Equation 11.6 and following steps described in Example 11.2.

(a) In these cases, Equation 11.6 changes to: $\dfrac{1}{d^2} = \dfrac{h^2 + k^2 + l^2}{a^2}$

The value of d for the (111) plane is then:

$$\frac{1}{d^2} = \frac{1^2 + 1^2 + 1^2}{a^2} = \frac{3}{(532\text{ pm})^2}$$

$$d = \sqrt{\frac{(532\text{ pm})^2}{3}} = 307\text{ pm}$$

The value of d for the (211) plane is:

$$\frac{1}{d^2} = \frac{2^2 + 1^2 + 1^2}{a^2} = \frac{6}{(532\text{ pm})^2}$$

$$d = \frac{532\text{ pm}}{\sqrt{6}} = 217\text{ pm}$$

And for the plane (100):

$$\frac{1}{d^2} = \frac{1^2}{a^2} = \frac{1}{(532\text{ pm})^2}$$

$$d = 532\text{ pm}$$

(b) In a similar way the value of d for a (123) plane of an orthorhombic cell with $a = 0.754$ nm, $b = 0.623$ nm and $c = 0.422$ nm is:

$$\frac{1}{d^2} = \frac{1^2}{(0.754\text{ nm})^2} + \frac{2^2}{(0.623\text{ nm})^2} + \frac{3^2}{(0.433\text{ nm})^2}$$

$$\frac{1}{d^2} = 60\text{ nm}^{-2}$$

$$d = 0.13\text{ nm}$$

And for the plane (236):

$$\frac{1}{d^2} = \frac{2^2}{(0.754\text{ nm})^2} + \frac{3^2}{(0.623\text{ nm})^2} + \frac{6^2}{(0.433\text{ nm})^2}$$

$$\frac{1}{d^2} = 222\text{ nm}^{-2}$$

$$d = 0.067\text{ nm}$$

11.23 Using the Bragg law:

$$\lambda = 2d\sin\theta$$

The wavelength is

$$\lambda = 2d\sin\theta = 2(97.3\text{ pm})\sin(19.85^\circ) = 66.1\text{ pm}$$

(In principle the reflection could be due to several wavelengths, such that

$$n\lambda = 66.1\text{ pm} \quad n = 1,2,3,\ldots$$

The wavelength we have given is the longest of this set.)

11.24 To construct the electron density along the x-axis we substitute the given structure factors into equation 11.8 yielding:

$V\rho(x) = 30 + (16.4\cos(2\pi x)) + (13.0\cos(4\pi x)) + (8.2\cos(6\pi x)) + (11\cos(8\pi x)) - 4.8\cos(10\pi x)) + (10.8\cos(12\pi x)) + (6.4\cos(14\pi x)) - (2.0\cos(16\pi x)) + (2.2\cos(18\pi x)) + (13\cos(20\pi x)) + (10.4\cos(22\pi x)) - (8.6\cos(24\pi x)) - (2.4\cos(26\pi x)) - (0.2\cos(28\pi x)) + (4.2\cos(30\pi x))$.

A graph of $V\rho(x)$ is shown below

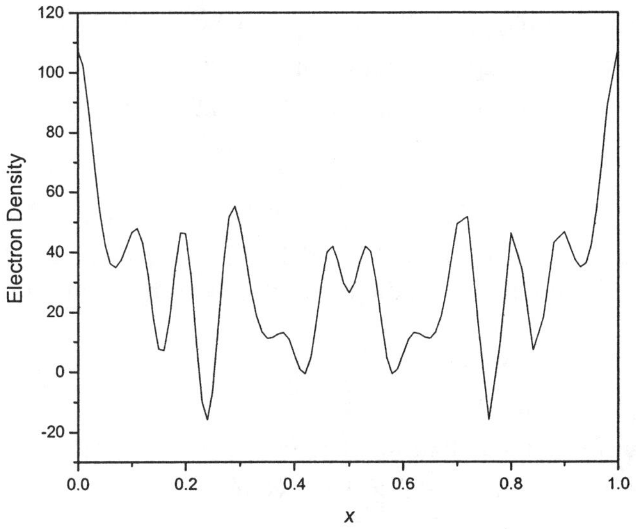

11.25 Assume a linear arrangement of the atoms forming the peptide hydrogen bond:

|← 290 pm →|
N---H........O====C
|←r→| |←R→|

In the electrostatic model of the hydrogen bond, we consider the electrostatic interactions of the three atoms: $N^{\delta-}$–$H^{\delta+}$...$O^{\delta-}$. We are given only the N...C distance (0.29 nm = 290 pm) and have to calculate the N...O and H...O distances. To find them, we calculate the N-H and C=O bond lengths using the atomic coordinates supplied in Structure (5):

$$r = [(182 - 132)^2 + (-87-0)^2]^{1/2} = 100 \text{ pm}$$

$$R = [(-62-0)^2 + (107-0)^2]^{1/2} = 124 \text{ pm}$$

(a) In a vacuum, $\varepsilon_r = 1$, the electrostatic interaction is

$$V = q_O q_H/(4\pi\varepsilon_0 r_{OH}) + q_O q_N/(4\pi\varepsilon_0 r_{ON}) = (e^2/4\pi\varepsilon_0)[\, z_O z_H/ r_{OH} + z_O z_N/ r_{ON}\,]$$

where we have introduced the fractional charges

$$q_O = z_O e \qquad q_N = z_N e \qquad q_H = z_H e$$

Based on Structure (5), the charges are

$$z_O = -0.38,\ z_H = +0.18 \text{ and } z_N = -0.36$$

e is the proton charge. Using our calculated bond lengths, we find

$$r_{OH} = 290 - 100 - 124 = 66 \text{ pm}$$

$$r_{ON} = 290 - 124 = 166 \text{ pm}$$

$$V = (e^2/4\pi\varepsilon_0)[z_O z_H/r_{OH} + z_O z_N/r_{ON}]$$

$= [(1.602\times10^{-19}\text{ C})^2/4\pi(8.854\times10^{-12}\text{ J}^{-1}\text{C}^2\text{m}^{-1})][-0.0684/(66\text{ pm}) + 0.1368/(166\text{ pm})]$

$= [2.307\times10^{-28}\text{J m}][-2.123\times10^{-4}\text{pm}^{-1}] = -4.9\times10^{-20}\text{ J} = -29\text{ kJ/mol}$

(b) In a lipid bilayer, $\varepsilon_r = 2$, the electrostatic interaction is

$$V = (e^2/4\pi\varepsilon_0\varepsilon_r)[z_O z_H/r_{OH} + z_O z_N/r_{ON}] = (-29\text{ kJ/mol})/2 = -14\text{ kJ/mol}$$

(c) In water, $\varepsilon_r \approx 80$, the electrostatic interaction is

$$V = (e^2/4\pi\varepsilon_0\varepsilon_r)[z_O z_H/r_{OH} + z_O z_N/r_{ON}] = (-29\text{ kJ/mol})/80$$

$= -0.36$ kJ/mol

11.26 The electronegativities of H and Cl are 2.1 and 3.0, respectively, according to Table 10.2. We could estimate the dipole moment of HCl by using Equation 11.10:

$$\mu/\text{D} \approx \Delta\chi$$

$$\mu \approx 0.9\text{ D}$$

The dipole moment of HCl in coulomb-meters is

$0.9\text{ D} \times 3.33564 \times 10^{-30}\text{ C m D}^{-1} = 3 \times 10^{-30}\text{ C m}$

11.27 (a) Use the formula provided

$$\mu_{res} = (\mu_1^2 + \mu_2^2 + 2\mu_1\mu_2\cos(\theta))^{1/2}$$

$= [(1.50)^2 + (0.80)^2 + 2(1.50)(0.80)\cos(109.5°)]^{½}\text{ D} = 1.44\text{ D}$

(b) In this case $\mu_1 = \mu_2 = \mu$

$$\mu_{res} = (\mu^2 + \mu^2 + 2\mu^2\cos(\theta))^{1/2} = \mu(2 + 2\cos(\theta))^{1/2}$$

For ortho substituents $\theta = 60°$

$$\mu_{ortho} = \mu(2 + 2\cos(\theta))^{1/2} = 3^{1/2}\mu \approx 1.73\mu$$

For meta substituents $\theta = 120°$

$$\mu_{meta} = \mu(2 + 2\cos(\theta))^{1/2} = \mu$$

The ratio:

$$\mu_{ortho}/\mu_{meta} = 3^{1/2} \approx 1.73$$

11.28 As described in Illustration 11.3, the electric dipole moment of glycine will be calculated by using Equation 11.11b to determine each of the components of the dipole moment, and then Equation 11.11a.

The structure of glycine below is showing partial charges on each atom using Table 11.2:

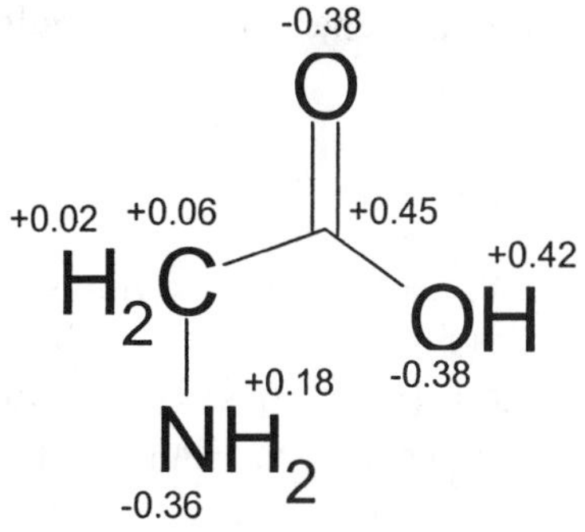

The expression for μ_x is:
$\mu_x = (0.18e) \times (-80 \text{ pm}) + (0.18e) \times (-199 \text{ pm}) + (-0.36e) \times (-101 \text{ pm}) + (0.02e) \times (-86 \text{ pm}) + (0.02e) \times (34 \text{ pm}) + (0.06e) \times (-195 \text{ pm}) + (0.42e) \times (129 \text{ pm}) + (0.45e) \times (82 \text{ pm}) + (-0.38e) \times (199 \text{ pm}) + (-0.38e) \times (49 \text{ pm})$
$\mu_x = -29.76e \text{ pm} = -29.76 \times (1.609 \times 10^{-19} \text{ C}) \times (10^{-12} \text{ m/pm})$
$= -4.788 \times 10^{-30} \text{ C m}$
or -1.43 D

The expression for μ_y is:
$\mu_y = (0.18e) \times (-110 \text{ pm}) + (0.18e) \times (-1 \text{ pm}) + (-0.36e) \times (-11 \text{ pm}) + (0.02e) \times (118 \text{ pm}) + (0.02e) \times (146 \text{ pm}) + (0.06e) \times (70 \text{ pm}) + (0.42e) \times (-146 \text{ pm}) + (0.45e) \times (-15\text{pm}) + (-0.38e) \times (16 \text{ pm}) + (-0.38e) \times (-107 \text{ pm})$
$\mu_y = -40.03e \text{ pm} = -40.03 \times (1.609 \times 10^{-19} \text{ C}) \times (10^{-12} \text{ m/pm})$
$= -6.441 \times 10^{-30} \text{ C m}$
or -1.93 D

The expression for μ_z is:
$\mu_z = (0.18e) \times (-111 \text{ pm}) + (0.18e) \times (-100 \text{ pm}) + (-0.36e) \times (-126 \text{ pm}) + (0.02e) \times (37 \text{ pm}) + (0.02e) \times (-98 \text{ pm}) + (0.06e) \times (-38 \text{ pm}) + (0.42e) \times (126 \text{ pm}) + (0.45e) \times (34\text{pm}) + (-0.38e) \times (-38 \text{ pm}) + (-0.38e) \times (88 \text{ pm})$
$\mu_z = 53.1e \text{ pm} = 53.1 \times (1.609 \times 10^{-19} \text{ C}) \times (10^{-12} \text{ m/pm})$
$= 8.54 \times 10^{-30} \text{ C m}$
or 2.56 D

Using Equation 11.11a, the magnitude of the dipole moment of glycine is:

$$\mu = \sqrt{(\mu_x^2 + \mu_y^2\ \mu_z^2)} = \sqrt{(-1.43 \text{ D})^2 + (-1.93 \text{ D})^2 + (2.56 \text{ D})^2} = 3.51 \text{ D}$$

11.29 (a) The molecular structure of hydrogen peroxide (HOOH) is shown in Structure 22. To analyze the charge distribution, consider that the two O atoms are equivalent and the two H atoms are equivalent, and that the whole molecule is electrically neutral. Due to this we must have:

charge on each H $q_H = ze$

charge on each O $q_O = -ze$

We choose a coordinate system in which calculations of the dipole moment will be especially easy: place O atoms on z axis with origin at the center of the molecule, and place one of the H atoms on the x axis (this is possible because the O—O—H angle is 90°). The O—H bond involving the second H atom will then lie parallel to the xy plane, forming an angle ϕ with the x axis. The (x,y,z) coordinates of the atoms will be:

O_1 : $(0,0,-a/2)$ $\quad a = 149$ pm is the O—O bond length,
O_2 : $(0,0,a/2)$
H_1 : $(b,0,-a/2)$ $\quad b = 97$ pm is the O—H bond length
H_2 : $(b\cos(\phi), b\sin(\phi), a/2)$

and the components of the dipole moment vector

$$\mu_x = 0(-ze) + 0(-ze) + b(ze) + b\cos(\phi)(ze) = zeb(1+\cos(\phi)) = \mu_0(1+\cos(\phi))$$
$$\mu_y = 0(-ze) + 0(-ze) + 0(ze) + b\sin(\phi)(ze) = zeb\sin(\phi) = \mu_0\sin(\phi)$$
$$\mu_z = (-a/2)(-ze) + (a/2)(-ze) + (-a/2)(ze) + (a/2)(ze) = 0$$

Where the quantity $zeb = \mu_0$ is the dipole moment of one O—H bond. The magnitude of the molecular dipole moment is thus

$$\mu = (\mu_x^2 + \mu_y^2 + \mu_z^2)^{1/2} = \mu_0[(1+\cos(\phi))^2 + (\sin(\phi))^2]^{1/2}$$
$$= \mu_0(2 + 2\cos(\phi))^{1/2} = \mu_0 2^{1/2}(1+\cos(\phi))^{1/2}$$

or,

$$\mu / \mu_0 = 2^{1/2}(1+\cos(\phi))^{1/2}$$

Variation of dipole moment of HOOH with angle phi

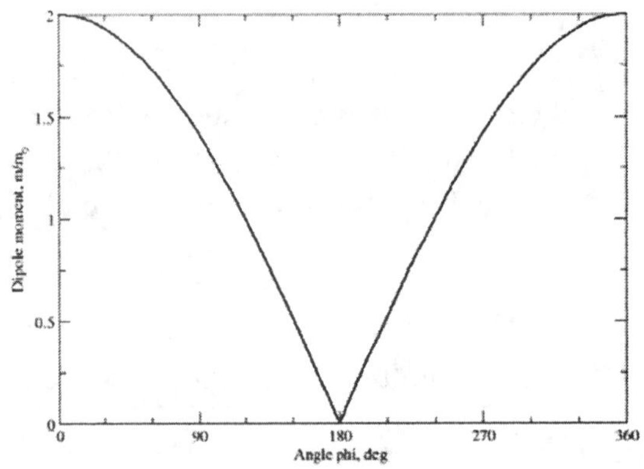

This quantity is plotted in the figure above. As might be expected, the dipole moment is largest, $\mu = 2\mu_0$, when the two O—H bond dipoles are aligned ($\phi = 0°$), and smallest, $\mu = 0$, when they point in opposite directions ($\phi = 180°$).

(b) One way to describe the orientation is by calculating the angle θ between the dipole moment vector and the x axis.

$$\tan(\theta) = \mu_y / \mu_x = \sin(\phi)/(1 + \cos(\phi))$$

$$\theta = \mathrm{atan}[\sin(\phi)/(1 + \cos(\phi))]$$

This interesting function is shown in the figure below. The angle θ is undefined for $\phi = 180^{\circ}$, when the dipole moment is zero.

Variation of orientation of HOOH dipole with angle phi

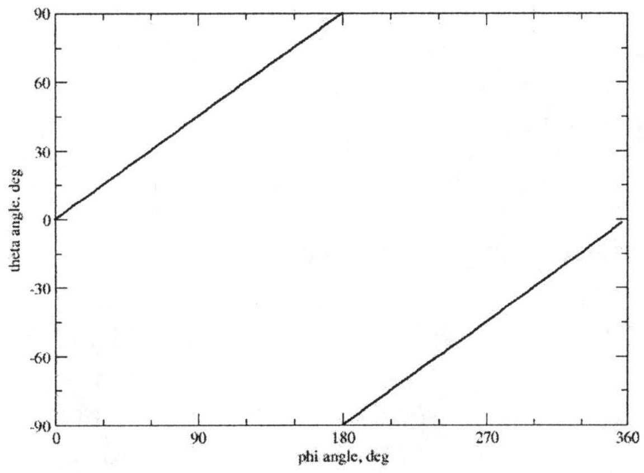

11.30 The molar energy required to reverse the direction of a water molecule would have to be twice the difference between the potential energy given by 11.12b with θ=0 and 180°.

(a) If the distance between the water molecule and Li^+ is 100 pm, the molar energy is:

$$V = 2\frac{q\mu}{4\pi\varepsilon_0 r^2} = \frac{1.602 \times 10^{-19}\ \mathrm{C} \times 1.85\ \mathrm{D} \times 3.33564 \times 10^{-30}\ \mathrm{C\,m\,D^{-1}}}{2 \times \pi \times 8.854 \times 10^{-12}\ \mathrm{J^{-1}\,C^2\,m^{-1}} \times \left(1.00 \times 10^{-10}\ \mathrm{m}\right)^2}$$

$V = 1.77 \times 10^{-18}$ J

After multiplication by Avogadro's number we get, $V = 1070\ \mathrm{kJ\,mol^{-1}}$

(b) If the distance between the water molecule and Li^+ is 300 pm the molar energy is:

$$V = 2\frac{q\mu}{4\pi\varepsilon_0 r^2} = \frac{1.602 \times 10^{-19}\ \mathrm{C} \times 1.85\ \mathrm{D} \times 3.33564 \times 10^{-30}\ \mathrm{C\,m\,D^{-1}}}{2 \times \pi \times 8.854 \times 10^{-12}\ \mathrm{J^{-1}\,C^2\,m^{-1}} \times \left(3.00 \times 10^{-10}\ \mathrm{m}\right)^2}$$

$V = 1.97 \times 10^{-19}$ J

After multiplication by Avogadro's number we get, $V = 119\ \mathrm{kJ\,mol^{-1}}$.

11.31 Structure 9 shows two parallel dipoles formed by four charges. We introduce a coordinate system in which the dipoles are aligned along the horizontal x axis and the y axis is vertical, as shown here:

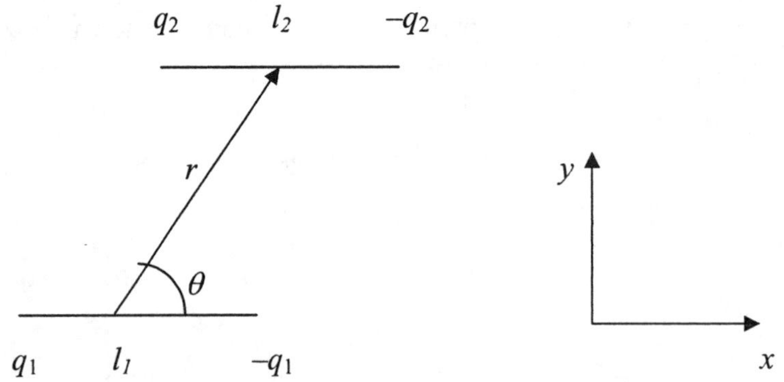

The atomic coordinates may be written as

atom	charge	x	y
1	q_1	$-l_1/2$	0
2	$-q_1$	$+l_1/2$	0
3	q_2	$-l_2/2 + r\cos(\theta)$	$r\sin(\theta)$
4	$-q_2$	$+l_2/2 + r\cos(\theta)$	$r\sin(\theta)$

The electrostatic energy is a sum of four terms

$$V = \frac{q_1q_2}{4\pi\varepsilon_0}\left[\frac{1}{\sqrt{\left(\left(\frac{l_1-l_2}{2}+r\cos\theta\right)^2+r^2\sin^2\theta\right)}}-\frac{1}{\sqrt{\left(\left(\frac{l_1+l_2}{2}+r\cos\theta\right)^2+r^2\sin^2\theta\right)}}\right]$$

$$+\frac{q_1q_2}{4\pi\varepsilon_0}\left[-\frac{1}{\sqrt{\left(\left(\frac{-l_1-l_2}{2}+r\cos\theta\right)^2+r^2\sin^2\theta\right)}}+\frac{1}{\sqrt{\left(\left(\frac{-l_1+l_2}{2}+r\cos\theta\right)^2+r^2\sin^2\theta\right)}}\right]$$

Expand the squares, add up $r^2\cos^2(\theta) + r^2\sin^2(\theta) = r^2$, and bring out a factor of r:

$$V = \frac{q_1q_2}{4\pi\varepsilon_0 r}\left[\frac{1}{\sqrt{1+\frac{(l_1-l_2)\cos\theta}{r}+\left(\frac{l_1-l_2}{2r}\right)^2}}-\frac{1}{\sqrt{1+\frac{(l_1+l_2)\cos\theta}{r}+\left(\frac{l_1+l_2}{2r}\right)^2}}\right]$$

$$+\frac{q_1q_2}{4\pi\varepsilon_0 r}\left[-\frac{1}{\sqrt{1+\frac{(-l_1-l_2)\cos\theta}{r}+\left(\frac{l_1+l_2}{2r}\right)^2}}+\frac{1}{\sqrt{1+\frac{(-l_1+l_2)\cos\theta}{r}+\left(\frac{l_1-l_2}{2r}\right)^2}}\right]$$

Remember that the Taylor expansion allows us to write

$$f(1+x) = f(1) + f'(1)x + (1/2)f''(1)x^2 + \ldots$$

For our function $f(x) = x^{-1/2}$, the derivatives are $f'(x) = -(1/2)\,x^{-3/2}$ and $f''(x) = (3/4)\,x^{-3/2}$ so

$$(1+x)^{-1/2} = 1 - (1/2)x + (3/8)x^2 + \ldots$$

Here is how it works for the first term in our energy, where the factor in the square bracket is the "x" from "$1+x$" and we write down the first two powers only:

$$(1 + (l_1 - l_2)\cos(\theta)/r + (l_1 - l_2)^2/4r^2)^{-1/2} \approx$$
$$= 1 - (1/2)[(l_1 - l_2)\cos(\theta)/r + (l_1 - l_2)^2/4r^2] + (3/8)[(l_1 - l_2)\cos(\theta)/r + (l_1 - l_2)^2/4r^2]^2$$

Our next step is to notice that our formula has terms with different powers of factors of type $(l_1 - l_2)/r$. For our dipole approximation to work, these terms have to be small, which we may write as

$$l_1/r << 1 \text{ and } l_2/r << 1$$

Physically, this means that the separation between the dipoles r is much greater than the dipole sizes (l_1, l_2 or any combination). Mathematically, the result is that succesive powers of factors looking like $(l_1 - l_2)/r$ become progressively smaller. For example, if

$$(l_1 - l_2)/r \approx 0.1, \text{ then } [(l_1 - l_2)/r]^2 \approx 0.01 \text{ and } [(l_1 - l_2)/r]^3 \approx 0.001.$$

Thus, we must find the first term in the powers of $(l_1 - l_2)/r$ that has a non-zero coefficient, and the following ones may be ignored. Based on Equation 11.13, we may suspect that we need to keep terms up to second order in $(l_1 - l_2)$/r, so from the third component of the expansion above we keep only the first part:

$$(3/8)[(l_1 - l_2)\cos(\theta)/r + (l_1 - l_2)^2/4r^2]^2 \approx 3(l_1 - l_2)^2 \cos^2(\theta)/8r^2$$

Finally, this leads to this expression for the first square root:

$$(1 + (l_1 - l_2)\cos(\theta)/r + (l_1 - l_2)^2/4r^2)^{-1/2} \approx$$
$$= 1 - (l_1 - l_2)\cos(\theta)/2r - (l_1 - l_2)^2/8r^2 + 3(l_1 - l_2)^2 \cos^2(\theta)/8r^2$$
$$= 1 - (l_1 - l_2)\cos(\theta)/2r - (1 - 3\cos^2(\theta))(l_1 - l_2)^2/8r^2$$

Following an analogous reasoning we can write for the three remaining square root expressions in our energy:

$$(1 + (l_1 + l_2)\cos(\theta)/r + (l_1 + l_2)^2/4r^2)^{-1/2} \approx$$
$$= 1 - (l_1 + l_2)\cos(\theta)/2r - (1 - 3\cos^2(\theta))(l_1 + l_2)^2/8r^2$$
$$(1 - (l_1 + l_2)\cos(\theta)/r + (l_1 + l_2)^2/4r^2)^{-1/2} \approx$$
$$= 1 + (l_1 + l_2)\cos(\theta)/2r - (1 - 3\cos^2(\theta))(l_1 + l_2)^2/8r^2$$
$$(1 + (-l_1 + l_2)\cos(\theta)/r + (l_1 - l_2)^2/4r^2)^{-1/2} \approx$$
$$= 1 + (l_1 - l_2)\cos(\theta)/2r - (1 - 3\cos^2(\theta))(l_1 - l_2)^2/8r^2$$

Now we sum up the terms with same powers of r, with the signs taken from our equation for V. For $1/r^0$

$$1 - 1 - 1 + 1 = 0$$

For $1/r^1$

$$[-(l_1 - l_2) + (l_1 + l_2) - (l_1 + l_2) + (l_1 - l_2)]\cos(\theta)/r = 0$$

For $1/r^2$

$$[-(l_1 - l_2)^2 + (l_1 + l_2)^2 + (l_1 + l_2)^2 - (l_1 - l_2)^2](1 - 3\cos^2(\theta))/8r^2$$
$$= l_1 l_2(1 - 3\cos^2(\theta))/r^2$$

The equation for electrostatic energy is thus

$$V = \frac{q_1 q_2}{4\pi\varepsilon_0 r^3} l_1 l_2 \left(1 - 3\cos^2\theta\right) = \frac{\mu_1 \mu_2 \left(1 - 3\cos^2\theta\right)}{4\pi\varepsilon_0 r^3}$$

when we identify $\mu_1 = q_1 l_1$ and $\mu_2 = q_2 l_2$.
The derivation of 11.13 is complete.

11.32 (a) The polarizability of a molecule, α, is defined as: μ^*/ε. The electric field defined as $\frac{q}{4\pi\varepsilon_0 r^2}$ has the following units: C/J^{-1} C^2 m^{-1} m^2

Therefore the units of polarizability are:

$$\alpha = \frac{\text{C m}}{\text{C/J}^{-1}\ \text{C}^2\ \text{m}^{-1}\ \text{m}^2}$$

$$\alpha = \text{J}^{-1}\ \text{C}^2\ \text{m}^2$$

(b) The polarizability volume α' is defined as: $\alpha/4\pi\varepsilon_0$. Therefore the units of polarizability volume are:

$$\alpha' = \frac{\alpha}{4\pi\varepsilon_0}$$

$$\alpha' = \frac{\text{J}^{-1}\ \text{C}^2\ \text{m}^2}{\text{J}^{-1}\ \text{C}^2\ \text{m}^{-1}}$$

$$\alpha' = \text{m}^3$$

11.33 The induced dipole moment μ^* is $\mu^* = \alpha E$, where α is the polarizability and E the electric field.
We are given the polarizability volume α', which is $\alpha' = \alpha/4\pi\varepsilon_0$.
For a water molecule at distance r from a proton ($q = e$) the induced dipole moment is

$$\mu^* = \alpha E = \alpha e/4\pi\varepsilon_0 r^2 = \alpha' e/r^2$$

so,

$$r = (\alpha' e/\mu^*)^{1/2} = [(1.48\times10^{-30}\text{m}^{-3})(1.602\times10^{-19}\text{C})/1.85(3.33564\times10^{-30}\text{C m})]^{1/2}$$
$$= 196\ \text{pm}$$

11.34 The energy of interaction between a benzene ring and peptide group is described by Equation 11.17:

$$V = -\frac{\mu_1^2 \alpha_2'}{4\pi\varepsilon_0 r^6} = -\frac{\left(2.7\ \text{D} \times 3.33564 \times 10^{-30}\ \text{C m D}^{-1}\right)^2 \times 1.04 \times 10^{-29}\ \text{m}^3}{4 \times \pi \times 8.854 \times 10^{-12}\ \text{J}^{-1}\ \text{C}^2\ \text{m}^{-1} \times \left(4.0 \times 10^{-9}\ \text{m}\right)^6}$$

$$V = -1.8 \times 10^{-27}\ \text{J}$$

11.35 The London formula for the dispersion interaction energy between two objects is Equation 11.18:

$$V = -C/R^6 \quad \text{with } C = (3/2)\alpha'_1\alpha'_2 I_1 I_2/(I_1 + I_2),$$

where α'_1 and α'_2 are the polarizability volumes, and I_1 and I_2 are the ionization potentials of the objects.

For two identical objects with $\alpha'_1 = \alpha'_2 = \alpha'$ and $I_1 = I_2 = I$:

$$C = (3/4)\alpha'^2 I$$

$$V = -3(1.04\times10^{-29}\text{m}^{-3})^2(4.8\times10^2\text{kJ mol}^{-1})/4(4.0\times10^{-9}\text{m})^6$$

$$= -9.5\times10^{-6}\ \text{kJ mol}^{-1}$$

This is a very weak effect due to the fast fall off of V with distance R.

11.36 This exercise can be solved in a similar manner as Exercise 11.15. Structure 24 shows partial charges on the atoms involved in hydrogen bonding.

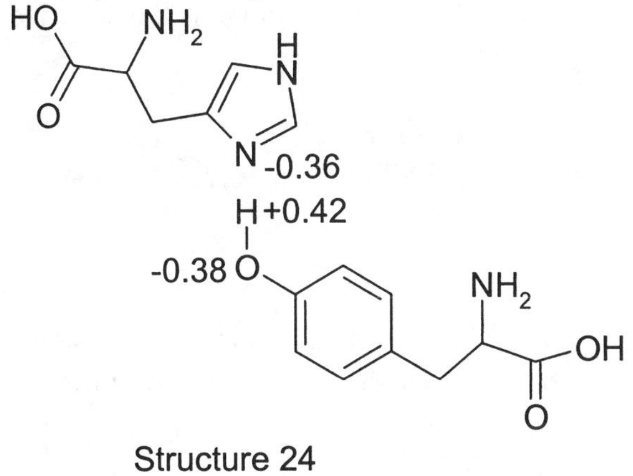

Structure 24

Assuming a linear geometry, the distance NH is given as 104.3 pm and OH will be taken as 97.5 pm. Then, the ON distance is 201.8 pm.

Using Equation 11.9.a for the electrostatic interaction between NH atoms:

$$V_{NH} = \frac{q_N q_H}{4\pi\varepsilon_0 r} = \frac{-0.36e \times 0.42e}{4 \times \pi \times 8.854 \times 10^{-12}\ \text{J}^{-1}\ \text{C}^2\ \text{m}^{-1} \times 1.043 \times 10^{-10}\ \text{m}}$$

$$V_{NH} = \frac{(-0.36 \times 0.42) \times (1.602 \times 10^{-19}\ \text{C})^2}{4 \times \pi \times 8.854 \times 10^{-12}\ \text{J}^{-1}\ \text{C}^2\ \text{m}^{-1} \times 1.043 \times 10^{-10}\ \text{m}}$$

$$V_{NH} = -3.344 \times 10^{-19}\ \text{J}$$

Multiplication by Avogadro's number we get: –201.3 kJ mol^{-1}
And the interaction between NO atoms:

$$V_{NO} = \frac{q_N q_O}{4\pi\varepsilon_0 r} = \frac{(-0.36e) \times (-0.38e)}{4 \times \pi \times 8.854 \times 10^{-12}\ \text{J}^{-1}\ \text{C}^2\ \text{m}^{-1} \times 2.018 \times 10^{-10}\ \text{m}}$$

$$V_{NO} = \frac{(0.36 \times 0.38) \times (1.602 \times 10^{-19}\ \text{C})^2}{4 \times \pi \times 8.854 \times 10^{-12}\ \text{J}^{-1}\ \text{C}^2\ \text{m}^{-1} \times 2.018 \times 10^{-10}\ \text{m}}$$

$$V_{NO} = 1.564 \times 10^{-19}\ \text{J}$$

Multiplication by Avogadro's number we get: 94.15 kJ mol^{-1}
The potential energy of the interaction is given by $V_{NH} + V_{NO}$, which in this case is: –107 kJ mol^{-1}.

11.37 The London dispersion interaction between two groups of atoms is

$$V = -C/R^6$$

with C depending on the type of groups involved. The force is the negative of the slope of the potential

$$F = -dV/dR \approx -(V(R+\delta R) - V(R))/\delta R$$

According to the expansion provided in the exercise,

$$V(R+\delta R) = -C/(R+\delta R)^6 = -C/[R^6(1+\delta R/R)^6] \approx -(C/R^6)(1 - 6\delta R/R)$$
$$= -C/R^6 + 6C\delta R/R^7$$

and

$$F \approx -6C/R^7$$

The force is negative for all values of R, indicating that this is an attractive interaction (i.e., increasing R increases V). This force will be zero for $R \to \infty$.

11.38 The differentiation of Equation 11.19 with respect to distance yields:

$$F = -\frac{dV}{dr} = \frac{6C}{r^7}$$

The force goes to zero as r approaches infinity.

11.39 The symmetric planar dimer of acetic acid (Structure 25) has a dipole moment of zero, because contributions from identical atoms of the two molecules cancel due to inversion symmetry. Acetic acid exists as a mixture of monomers and dimers, with the equilibrium

$$A + A \rightarrow A_2$$

described by the equilibrium constant *K*. The fact that the apparent dipole moment of gaseous acetic acid increases with increasing temperature may be explained by the decrease in the concentration of dimers (which have zero dipole moment), and corresponding increase of concentration of monomers (which have a non-zero dipole moment). This means that *K* decreases with increasing temperature, i.e., the enthalpy of dimer formation is negative. The dimer is more stable than two separate molecules (see Chemical Equilibrium).

11.40 This exercise is similar to Exercise 11.36, but the arrangement of atoms is not linear.

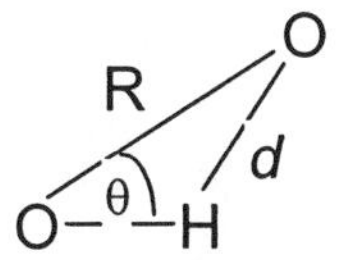

We are going to evaluate the electrostatic interaction between OO and OH atoms. The distance between OO atoms is given as R = 200 pm, but the OH distance (*d*) has to be calculated using Figure A2.9.
Thus,

$$d = (R^2 + r^2 - 2Rr\cos\theta)^{1/2}$$

Using Equation 11.9.a for the electrostatic interaction between OO atoms in kJ mol^{-1}:

$$V_{OO} = \frac{N_A q_O q_O}{4\pi\varepsilon_0 R} = \frac{6.02214 \times 10^{23}\,\text{mol}^{-1} \times (-0.83e)^2}{4 \times \pi \times 8.854 \times 10^{-12}\ \text{J}^{-1}\ \text{C}^2\ \text{m}^{-1} \times 2.00 \times 10^{-10}\ \text{m}}$$

$$V_{OO} = \frac{6.02214 \times 10^{23}\,\text{mol}^{-1} \times (0.83)^2 \times (1.602 \times 10^{-19}\ \text{C})^2}{4 \times \pi \times 8.854 \times 10^{-12}\ \text{J}^{-1}\ \text{C}^2\ \text{m}^{-1} \times 2.00 \times 10^{-10}\ \text{m}}$$

$$V_{OO} = 478.47\ \text{kJ mol}^{-1}$$

And the interaction between OH atoms:

$$V_{OH} = \frac{N_A q_O q_H}{4\pi\varepsilon_0 R} = \frac{6.02214 \times 10^{23}\ \text{mol}^{-1} \times (-0.83e) \times (0.45e)}{4 \times \pi \times 8.854 \times 10^{-12}\ \text{J}^{-1}\ \text{C}^2\ \text{m}^{-1} \times \sqrt{(R^2 + r^2 - 2Rr\cos\theta)}\ \text{m}}$$

$$V_{OH} = \frac{N_A q_O q_H}{4\pi\varepsilon_0 R} = -\frac{6.02214 \times 10^{23}\ \text{mol}^{-1} \times (0.373) \times (1.602 \times 10^{-19}\ \text{C})^2}{4 \times \pi \times 8.854 \times 10^{-12}\ \text{J}^{-1}\ \text{C}^2\ \text{m}^{-1} \times \sqrt{(R^2 + r^2 - 2Rr\cos\theta)}\ \text{m}}$$

$$V_{OH} = \frac{N_A q_O q_H}{4\pi\varepsilon_0 R} = -\frac{5.181 \times 10^{-5}}{\sqrt{(4.9116 \times 10^{-20} - 3.828 \times 10^{-20} \times \cos\theta)}}\ \text{J mol}^{-1}$$

The total molar potential energy, V, is $V_{OO} + V_{OH}$ and a graph or V versus θ is shown below:

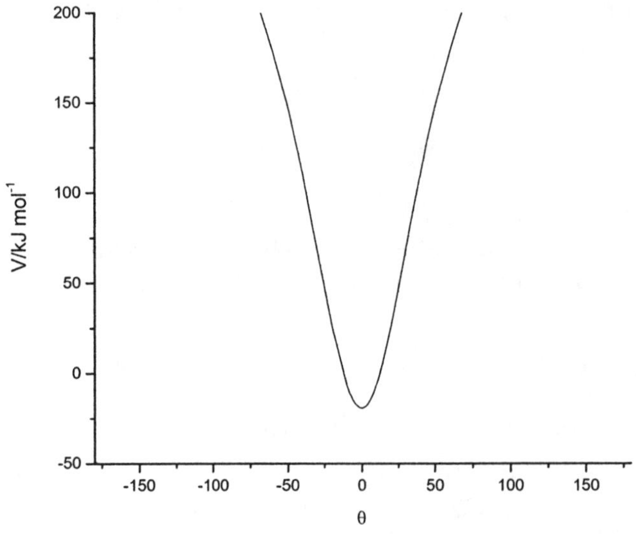

11.41 As discussed in Section 11.13, there are two reasons why parallel *β*-sheets are not common in proteins. First, the N—H…O atoms in the hydrogen bonds linking the peptide strands are not perfectly aligned, which makes the interaction less favorable (see Exercise 11.40).
Second, the N—H bonds on neighboring chains are aligned with each other, as are the C=O bonds. Treating the bonds as dipoles, this is a relative orientation of $\theta = 90^\circ$ in Equation 11.13, which corresponds to a positive, unfavorable, interaction energy (Exercise 11.29).

11.42 (a) The fully eclipsed conformation is achieved at $\phi = 0$ or $\cos 3\phi = 1$. Therefore, the change in potential energy between the *trans* and fully eclipsed conformations is 11.6 kJ mol^{-1}.

The potential energy as a function of the azimuthal angle is shown in the graph below:

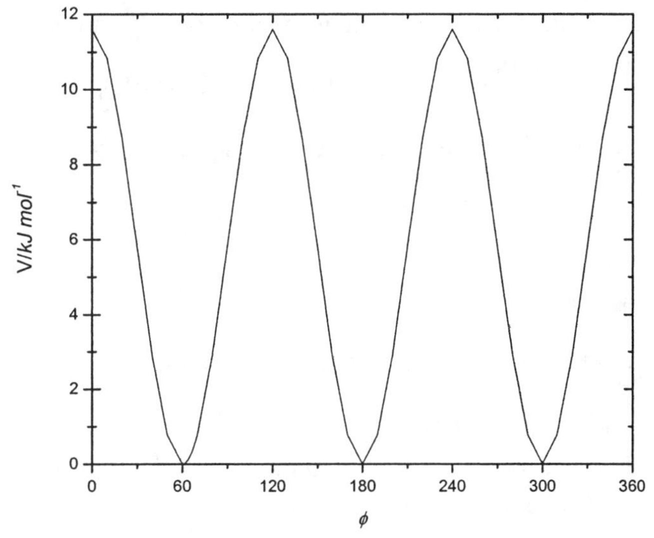

(b) We use the Taylor expansion series for $\cos x = 1 - \dfrac{x^2}{2!} + \ldots$

The potential energy is now written as:

$$V = \frac{1}{2}V_0\left(1 + 1 - \frac{(3\phi)^2}{2}\right) = V_0 - \frac{V_0}{4}(3\phi)^2$$

Or,

$$V = V_0\left(1 - \frac{1}{4}(3\phi)^2\right)$$

This expression can be identified with Equation 9.18b.

(c) Assuming the torsion energy for small variations in angle around the C-C bond is 1 kJ mol^{-1}, the vibrational frequency is then:

$$\nu = \frac{1.66 \times 10^{-21}\ \mathrm{J}}{6.626 \times 10^{-34}\ \mathrm{J\ s}} = 2.5 \times 10^{12}\ \mathrm{s}^{-1}$$

11.43 For a flexible chain, the relation between the r.m.s. separation of ends R_{ms}, the number of links N and length of a single link l is

$$R_{ms} = N^{1/2}l = (700)^{1/2}(0.90\ \mathrm{nm}) = 23.8\ \mathrm{nm}.$$

11.44 (a) From Equation 11.25, we can calculate the contour length, R_c. The number of residues N is obtained by dividing the molar mass of the macromolecule by the molar mass of each residue CH_2-CH_2. The length of each residue, l, is calculated by adding the bond length of C-C to the distance of a carbon atom radius at each end of the monomer. In other words, l is equal to 2C-C bond lengths.

Solving for $R_c = \frac{280000 \text{ g mol}^{-1}}{28 \text{ g mol}^{-1}} \times 2 \times 154 \text{ pm} = 3.08 \times 10^6 \text{ pm}$

(b) From Equation 11.24 we calculate the root mean square separation:

$R_{rms} = N^{1/2} l = \sqrt{1.00 \times 10^4} \times 2 \times 154 \text{ pm}$

$R_{rms} = 3.08 \times 10^4 \text{ pm}$

11.45 For a polymer in the random coil state, the radius of gyration is

$$R_g = (N/6)^{1/2} l$$

where N is the number of links and l the link length.

The length of a single C—C bond is about 0.15 nm. Thus, the number of links is

$$N = 6(R_g/l)^2 = 6(7.3 \text{ nm}/0.15 \text{ nm})^2 = 1.4 \times 10^4.$$

11.46 The drawing below represents a random walk:

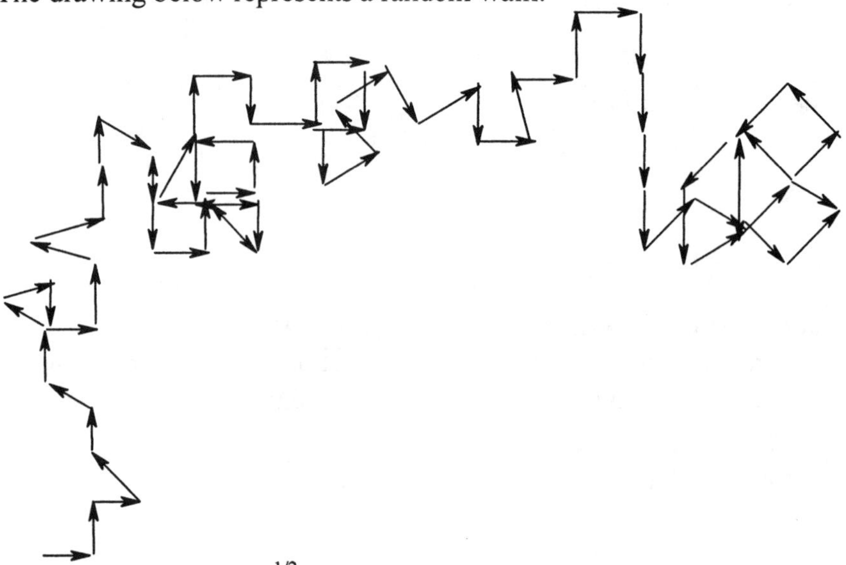

It seems they vary as $N^{1/2}$.

11.47 (a) For a molecule of radius R, the molecular volume is

$$v = 4\pi R^3/3$$

The molar volume is then

$$V_m = N_A v = 4\pi N_A R^3/3$$

with the N_A the Avogadro constant. Now let V, m be the total volume and mass of sample and n be the amount of substance. The specific volume is the inverse of the mass density:

$$v_s = V/m = (nV_m)/(nM) = V_m/M \text{ or } V_m = Mv_s$$

where M is the molar mass. For a spherical molecule of radius R, the radius of gyration is:

$$R_g = (3/5)^{½}R = (3/5)^{½}(3V_m/4\pi N_A)^{1/3} = (3/5)^{½}(3Mv_s/4\pi N_A)^{1/3}$$

if we express v_s in cm^3/g, M in g/mol and N_A in 1/mol, the units of R_g will be 1 cm = 10^7 nm. Thus,

$$R_g\text{/nm} = (3/5)^{½}R = (3/5)^{½}(3/4\pi\times 6.022\times 10^{23})^{1/3}\,[(v_s/\text{cm}^3\text{g}^{-1})(M/\text{g mol}^{-1})]^{1/3}(10^7)$$

$$R_g\text{/nm} = 0.0569022\times[(v_s/\text{cm}^3\text{g}^{-1})(M/\text{g mol}^{-1})]^{1/3}$$

which proves the relation given in this Exercise.

(b) The radii of gyration corresponding to a solid sphere are

serum albumin: $R_g^0 = 0.057\times(66\times 10^3\times 0.752)^{1/3} = 2.1$ nm

bushy stunt virus: $R_g^0 = 0.057\times(10.6\times 10^6\times 0.741)^{1/3} = 11$ nm

DNA: $R_g^0 = 0.057\times(4\times 10^6\times 0.556)^{1/3} = 7.4$ nm

For serum albumin and the bushy stunt virus the calculated values are similar to the actual values, so these molecules are approximately spherical. For DNA the actual radius of gyration is 117 nm, much larger than expected for a spherical object of the same mass (7.4 nm). We conclude that the DNA is not spherical – it could be rod-like in shape.

11.48 (a) The exponential-6 potential is:

$$V = 4\varepsilon\left[e^{-r/\sigma} - \left(\frac{\sigma}{r}\right)^6\right]$$

A graph of V versus r is shown below:

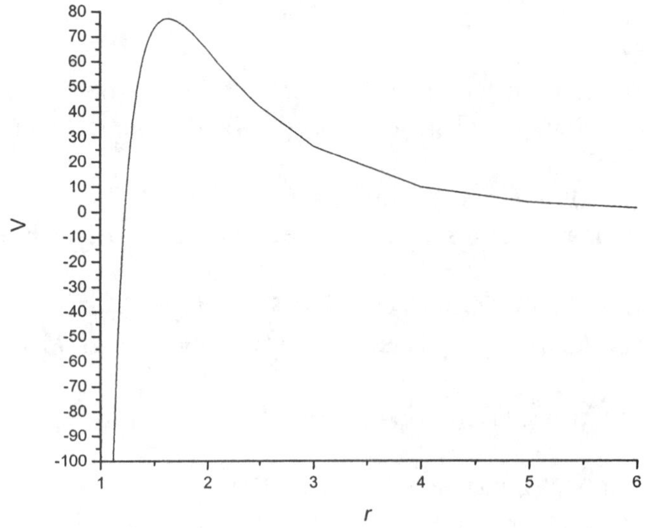

(b) The minimum is found, as shown in Self-Test 11.10, by setting the dV/dr = 0

$$\frac{dV}{dr} = 4\varepsilon\left(-\frac{e^{-r/\sigma}}{\sigma} + 6\frac{\sigma^6}{r^7}\right) = 0$$

$$\frac{e^{-r/\sigma}}{\sigma} = 6\frac{\sigma^6}{r^7}$$

$$e^{-r/\sigma} = 6\frac{\sigma^7}{r^7}$$

11.49 (a) Fitting the data to the equation

$$\log A = b_0 + b_1 S + b_2 W$$

gives

$$b_0 = 3.5903$$
$$b_1 = 0.9571$$
$$b_2 = 0.3619$$

(b) Given the equation from part (a), we can try to predict the value of W:

$$W = (\log A - b_0 - b_1 S)/b_2 = (7.60 - 3.5903 - 0.9571\times4.84) = -1.72$$

This value does not appear reasonable, as all other W values used to determine the model equation, were between 1.00 and 1.95. The log$A$ and S values for this compound do not fit our model. This might be because it represents a new class of molecules for which we should have to re-fit our model equation.

Chapter 12: Statistical Aspects of Structure and Change

12.9 The binomial coefficients C(4,n)are given by the expansion of $(1+x)^4$ = 1,4,6,4,1. Thus, for 4 binding sites and 0 to 4 ligands binding to these sites, we get the following combinations. (Note: L1L2 = ligand to site #1 and a ligand to site 2, etc.)

Number of ligands	Possible Arrangements	Total
0	1	1
1	L1,L2,L3,L4	4
2	L1L2, L1L3, L1L4, L2L3, L2L4, L3L4	6
3	L1L2L3,L1L2L4, L1L3L4, L2L3L4	4
4	L1234	1

12.10 (a) In Derivation 12.2, Stirling's approximation was used in the form ln(2 π) + (x + ½) ln − x.

If a cruder approximation $N! = N^N$ is used, then ln $N!$ = N ln$N$

$$\ln P = N\ln N - \left[\frac{1}{2}(N+n)\right]\ln\left[\frac{1}{2}(N+n)\right] - \left[\frac{1}{2}(N-n)\right]\ln\left[\frac{1}{2}(N-n)\right] - N\ln 2$$

Using the laws of logarithms,

$$\ln P = N\ln\left(\frac{N}{2}\right) - \left[\frac{1}{2}(N+n)\right]\ln\left[\frac{1}{2}(N+n)\right] - \left[\frac{1}{2}(N-n)\right]\ln\left[\frac{1}{2}(N-n)\right]$$

(b) The precise form of Stirling's approximation is $x! = (2\pi)^{1/2}\, x^{x+1/2}\, e^{-x}$ and ln $x!$ = ln(2 π) + (x + ½) ln − x, which is the one used in Derivation 12.2

12.11

$$\ln P = \left(N+\frac{1}{2}\right)\ln N - N - \left[\frac{1}{2}(N+n)+\frac{1}{2}\right]\ln\left[\frac{1}{2}(N+n)\right]+\frac{1}{2}(N+n) -$$
$$\left[\frac{1}{2}(N-n)+\frac{1}{2}\right]\ln\left[\frac{1}{2}(N-n)\right]+\frac{1}{2}(N-n)+\frac{1}{2}\ln 2\pi-\frac{1}{2}\ln 2\pi-\frac{1}{2}\ln 2\pi - N\ln 2$$
$$= \left(N+\frac{1}{2}\right)\ln N - N - \frac{1}{2}(N+n+1)\ln\left[\frac{1}{2}N\left(1+\frac{n}{N}\right)\right]+\frac{1}{2}(N+n) -$$
$$\frac{1}{2}(N-n+1)\ln\left[\frac{1}{2}N\left(1-\frac{n}{N}\right)\right]+\frac{1}{2}(N-n)+\frac{1}{2}\ln 2\pi-\frac{1}{2}\ln 2\pi-\frac{1}{2}\ln 2\pi - N\ln 2$$

Collecting terms and using the laws of logarithms gives,

$$\left(N+\frac{1}{2}\right)\ln N - \frac{1}{2}\ln 2\pi - N\ln 2 - \frac{1}{2}(N+n+1))\ln\left(1+\frac{n}{N}\right) -$$
$$\frac{1}{2}(N-n+1)\ln\left(1-\frac{n}{N}\right) - (N+1)\ln\frac{N}{2}$$
$$= \ln\left(\frac{N}{2\pi}\right)^{\frac{1}{2}} - \ln\left(\frac{N}{2}\right) - \frac{1}{2}(N+n+1))\ln\left(1+\frac{n}{N}\right) - \frac{1}{2}(N-n+1)\ln\left(1-\frac{n}{N}\right)$$
$$= \ln\left(\frac{2}{\pi N}\right)^{\frac{1}{2}} - \frac{1}{2}(N+n+1))\ln\left(1+\frac{n}{N}\right) - \frac{1}{2}(N-n+1)\ln\left(1-\frac{n}{N}\right)$$

which is the desired result.

12.12 The graph below shows how the probability changes with both time (100 and 885 ps) and the variable x. The step size was chosen as 200 pm, and the step time equal to 8.85 ps. The step size and step time are the values used in Figure 12.2.

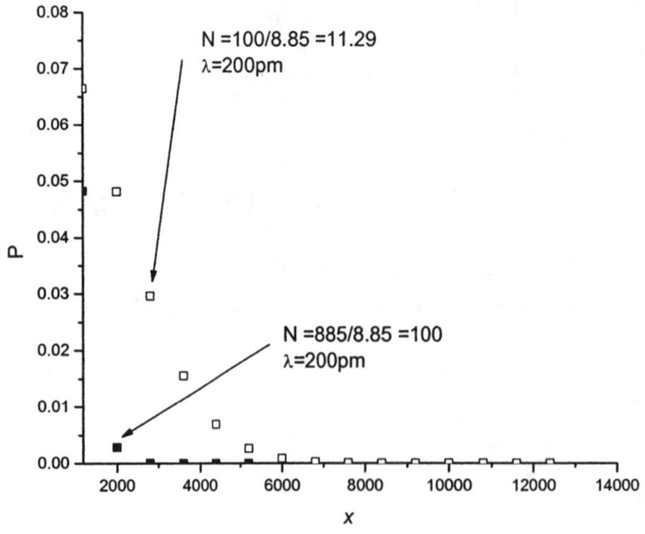

If the number of steps $N = t/\tau \rightarrow \infty$ the ratio $n^2/2N \rightarrow 0$, then the function (Equation 12.5)

$$P = \left(\frac{2}{\pi N}\right)^{1/2} e^{-n^2/2N},$$

in which the definitions of N and $x = n\lambda$ were included.
These can be changed by writing the exponential as an infinite series (see Appendix 2, Section A2.5)

$$e^{-n^2/2N} = 1 - \frac{n^2}{2N}$$

and the function changes to:

$$P = \left(\frac{2}{\pi N}\right)^{1/2}\left(1 - \frac{n^2}{2N}\right) \text{ as } N \rightarrow \infty$$

The graph below shows how the exponential function and $1 - \dfrac{n^2}{2N}$ changes with n at N=6000.

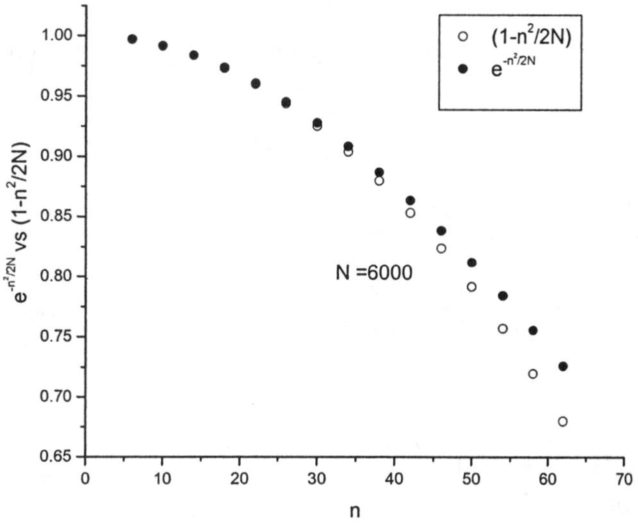

The two functions change with both N and n. Both of them are identical for a wide range of n if N is sufficiently high.

The graph below is showing the percent discrepancy between the two functions at different N values.

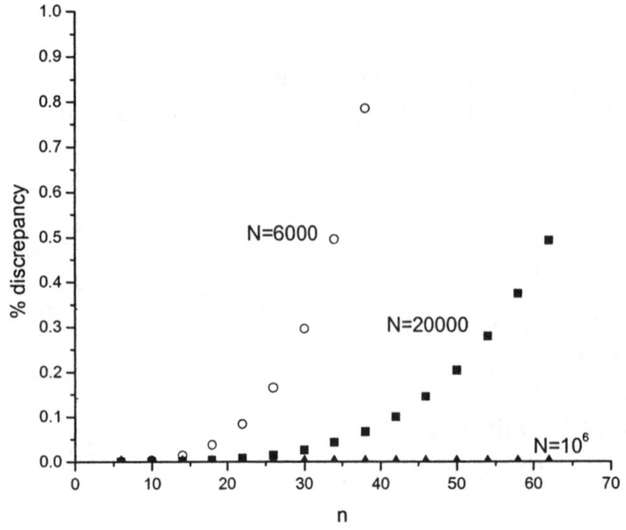

A percent discrepancy of 0.1 is obtained at higher values of n as N increases.

12.13

$$V = 100\frac{\text{kg}}{\text{mol}} \times \frac{1\text{ mol}}{6.022\times10^{23}} \times \frac{1\text{ m}^3}{1\text{ kg}} = 1.661\times10^{-22}\text{ m}^3 = \frac{4}{3}\pi a^3. \quad \therefore a = 3.41\times10^{-8}\text{ m}.$$

$$D = \frac{kT}{6\pi\eta a} = \frac{(1.381\times10^{-23}\text{ J/K})\times(298.15\text{ K})}{6\pi\times(1\times10^{-5}\text{ kg m}^{-1}\text{s}^{-1})\times(3.41\times10^{-8}\text{ m})} = 6.40\times10^{-10}\, m^2 s^{-1}$$

Then from

$$\tau = \frac{<\lambda^2>}{2D} = \frac{(1.0\text{ m})^2}{2(6.40\times10^{-10}\text{ m}^2\text{ s}^{-1})} = 7.8\times10^8\text{ s}.$$

Thus, diffusion is not the mechanism.

12.14 (a) If all the molecules have equal energies, there will be 5 molecules at the energy level 1ε. From Equation 12.4, we calculate W as:

$$W = \frac{5!}{0!5!0!0!0!} = 1$$

(b) The weights of different configurations are shown in the table below:

0ε	1ε	2ε	3ε	4ε	5ε	W
1	3	1	0	0	0	20
3	0	1	1	0	0	20
0	5	0	0	0	0	1
4	0	0	0	0	1	5
2	2	0	1	0	0	30
2	1	2	0	0	0	30
3	1	0	0	1	0	20

The most probable configurations are:

2	2	0	1	0	0

And

2*	1	2	0	0	0

12.15 (a)The constraints indicate that the sum of all states = 9ε and that $N = 9$. Thus, it is seen that putting molecules into the upper states limits the number of configurations. For instance, putting 1 in level 9 = 9ε puts 8 in level 0 = 0ε, which results in 9 possible configurations. We will therefore illustrate this point using a set of three.
Consider the following configurations labeled A,B and C with the indicated number of molecules.

Energy/ε	A	B	C
3	3	2	1
2	0	1	2
1	0	1	2
0	6	5	4

(b) It is obvious that Configuration C more evenly distributes the molecules, we therefore guess that it is the most probable.
(c) Calculation gives A = 9!/(3!0!0!6!) = 84, B = 9!/(2!1!1!5!) = 1,512 and C = 9!/(1!2!2!4!) = 22,680, which confirms our guess.

12.16 (a) From Equation 12.12 we have:

$$\frac{n_2}{n_1} = e^{-(\varepsilon_2 - \varepsilon_1)/kT} = e^{-\varepsilon j/kT},$$

if p_j is the most probable fraction of molecules in the stage with energy $j\varepsilon$:
We have

$$n_j = n_1\, e^{-\frac{\varepsilon j}{kT}}$$

Taking logarithms of this expression we get:

$$\ln n_j = \ln n_1 - \frac{\varepsilon j}{kT},$$

or $\ln p_j = \text{constant} - \dfrac{\varepsilon j}{k\mathrm{T}}$.

A graph of $\ln p_j$ against j or $\ln n_j$ against j, should provide T by determining the slope of the graph $\left(\dfrac{-\varepsilon}{k\mathrm{T}}\right)$.

The most probable configuration from Exercise 12.15 is 4221000000, and the corresponding $\ln p_j$ and j data is collected in the table below:

j	0	1	2	3
n_j	4	2	2	1
$\ln n_j$	1.386	0.693	0.693	0

A graph of $\ln n_j$ against j is shown below:

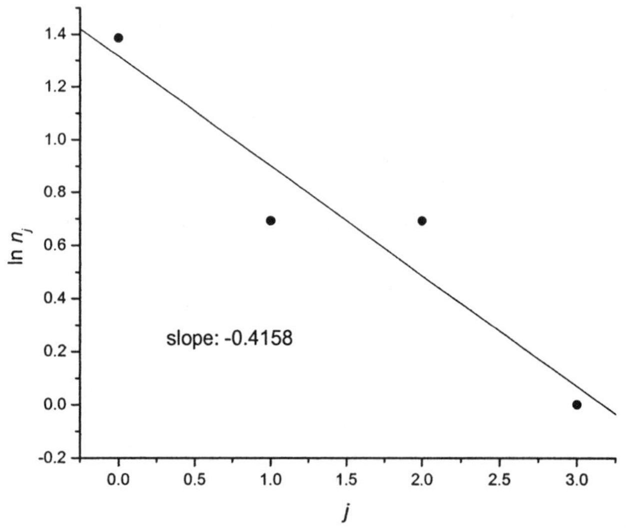

The slope of the graph is -0.4158 $= \dfrac{-\varepsilon}{k\mathrm{T}}$. The value of ε/hc is 50 cm^{-1}, and the temperature of the system is:

$$\mathrm{T} = \frac{\varepsilon}{k \times \text{slope}} = \frac{50\ \text{cm}^{-1} \times 2.998 \times 10^{10}\ \text{cm s}^{-1} \times 6.626 \times 10^{-34}\ \text{J s}}{1.381 \times 10^{-23}\ \text{J K}^{-1} \times 0.4158}$$

T = 173 K

(b) We could use the same procedure to configuration B from Exercise 12.15. The relevant data is shown below:

J	0	1	2	3
n_j	5	1	1	2
$\ln n_j$	1.609	0	0	0.693

A graph of ln n_j against j is shown below:

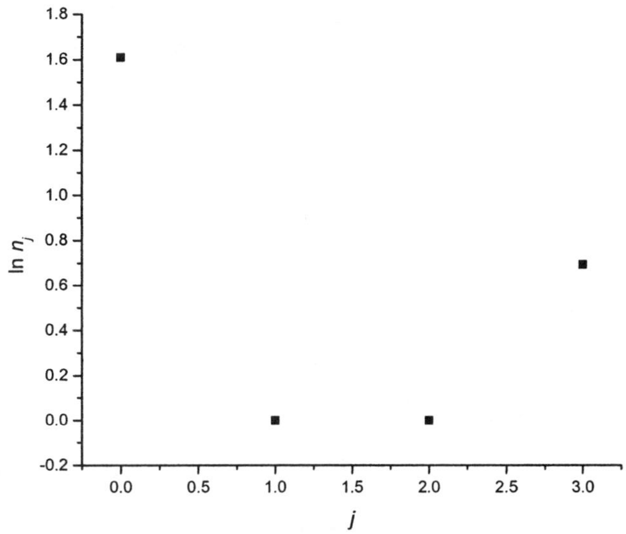

Clearly this configuration gives a poor correlation.

12.17

$$\frac{N_{stretched}}{N_{coiled}} = e^{-\left(\frac{\Delta\varepsilon}{kT}\right)} = e^{-\left(\frac{2.4\times10^{3}\,\text{J mol}^{-1}}{(8.314\,\text{J K}^{-1}\text{mol}^{-1})\times(293.15\,\text{K})}\right)} = 0.373$$

12.18 The partition function can be written by using Equation 12.11 and following Example 12.1. Setting the lower energy level to zero and noting that
$\Delta E = +\mu_B B - (-\mu_B B) = 2\,\mu_B B$
Thus for this case:
$q = 1 + e^{-2\mu_B B/kT}$

A graph of q against B is shown below:

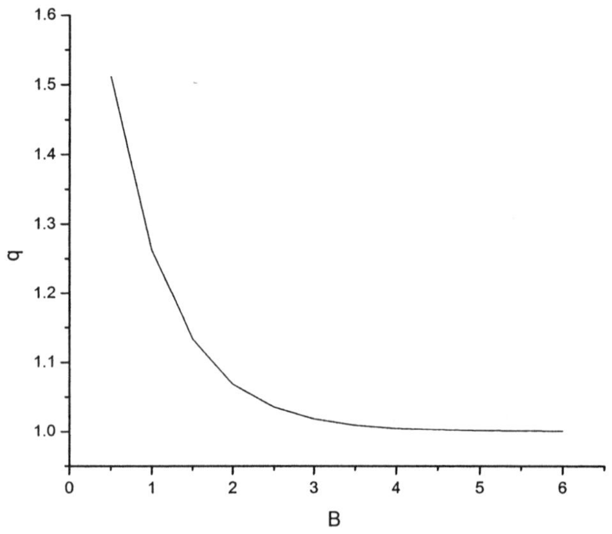

The exponent was calculated as follows:

$$2\mu_B B/kT = \frac{2 \times 9.274 \times 10^{-24}\ \text{J T}^{-1} \times \text{B}}{1.381 \times 10^{-23}\ \text{J K}^{-1} \times \text{T}} = 1.343\text{B/K}$$

(b) From Equation 12.12 we can calculate the relative populations at different T values:

At T = 4.0 K and B = 1T

$$\frac{n_2}{n_1} = \mathrm{e}^{-(\varepsilon_2 - \varepsilon_1)/kT} = \mathrm{e}^{\frac{1.343}{4.0}} = 0.71$$

At 298 K and B = 1T

$$\frac{n_2}{n_1} = \mathrm{e}^{-(\varepsilon_2 - \varepsilon_1)/kT} = \mathrm{e}^{\frac{1.343}{298.0}} = 0.996$$

12.19 (a) $q = \sum_i g_i \mathrm{e}^{-\beta\varepsilon_i} = 1 + 5\mathrm{e}^{-\beta\varepsilon} + 3\mathrm{e}^{-3\beta\varepsilon}$

(b) $q(T=0) = 1$ since only the first term contributes.

(c) $q(T=\infty) = 1+5+3 = 9$.

12.20 (a) The strategy will be to substitute Equation 12.13a for the partition function into the given equation:

First we will find dq/dT:

$$\frac{dq}{dT} = \frac{d}{dT}\left(\frac{1}{1 - e^{-h\nu/kT}}\right)$$

$$\frac{dq}{dT} = \frac{h\nu / k e^{-h\nu/kT}}{\left(1 - e^{-h\nu/kT}\right)^2 T^2}$$

The term $\frac{1}{\left(1 - e^{-h\nu/kT}\right)^2}$ is identified as q^2, then the above expression changes to:

$$\frac{dq}{dT} = \frac{q^2 h\nu}{kT^2} e^{-h\nu/kT}$$

Substitute dq/dT into the given expression for the mean energy yields:

$$\langle \varepsilon^M \rangle = \frac{kT^2}{q}\frac{dq}{dT} = \frac{kT^2}{q} \times \frac{q^2 h\nu}{kT^2} e^{-h\nu/kT}$$

$$\langle \varepsilon^M \rangle = qh\nu e^{-h\nu/kT}$$

Substitute Equation12.13a for the partition function:

$$\langle \varepsilon^M \rangle = \frac{h\nu}{\left(1 - e^{-h\nu/kT}\right)} e^{-h\nu/kT}$$

Divide the numerator and denominator by $e^{-h\nu/kT}$

$$\langle \varepsilon^M \rangle = \frac{h\nu}{\left(e^{h\nu/kT} - 1\right)}$$

(b) A graph of the expression above against temperature is shown below:

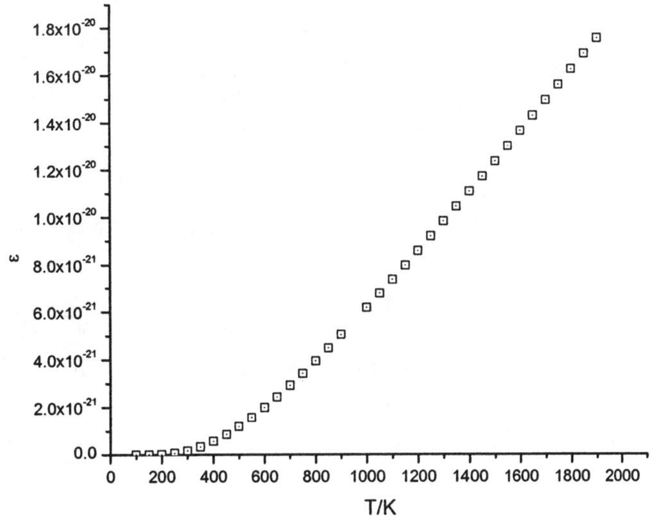

(c) At higher temperatures, the expression changes by writing the exponential as an infinite series (see Appendix 2, Section A2.5).

$$e^{h\nu/kT} = 1 + \frac{h\nu}{kT} + \ldots$$

Therefore, the expression in part (a) changes to:

$$\langle \varepsilon^M \rangle = \frac{h\nu}{\left(1 + \frac{h\nu}{kT} - 1\right)}$$

$$\langle \varepsilon^M \rangle = kT$$

12.21

$$U = U(0) - N\left(\frac{\mathrm{d}\ln q}{\mathrm{d}\beta}\right) = U(0) - \frac{N}{q}\left(\frac{\partial q}{\partial \beta}\right) \qquad q_{\text{vib}} = \frac{1}{1-\mathrm{e}^{-h\nu/kT}} = \frac{1}{1-\mathrm{e}^{-\beta h\nu}}$$

$$\frac{\mathrm{d}q}{\mathrm{d}\beta} = \frac{\mathrm{d}}{\mathrm{d}\beta}(1-\mathrm{e}^{-h\nu\beta})^{-1} = \frac{-h\nu\mathrm{e}^{-h\nu\beta}}{(1-\mathrm{e}^{-h\nu\beta})^2}$$

$$U = U(0) - N\left(\frac{\mathrm{d}\ln q}{\mathrm{d}\beta}\right) = U(0) - \frac{N}{q}\left(\frac{\partial q}{\partial \beta}\right) = \frac{1}{2}Nh\nu + \frac{Nh\nu\mathrm{e}^{-h\nu\beta}}{1-\mathrm{e}^{-h\nu\beta}}$$

12.22 The rotational partition function is calculated using Equation 12.15a.

(a) For $^{1}H^{35}Cl$, σ is equal to 1 since HCl is an unsymmetrical linear rotor and q is:

$$q = \frac{kT}{\sigma hB} = \frac{1.381 \times 10^{-23}\ \mathrm{J\,K^{-1}} \times 298\ \mathrm{K}}{1 \times 6.626 \times 10^{-34}\ \mathrm{J\,s} \times 318 \times 10^{9}\ \mathrm{s^{-1}}}$$

$$q = 19.5$$

(b) For $^{12}C^{16}O_2$, σ is equal to 2 since CO_2 is a symmetrical linear rotor and q is:

$$q = \frac{kT}{\sigma hB} = \frac{1.381 \times 10^{-23}\ \mathrm{J\,K^{-1}} \times 298\ \mathrm{K}}{2 \times 6.626 \times 10^{-34}\ \mathrm{J\,s} \times 11.70 \times 10^{9}\ \mathrm{s^{-1}}}$$

$$q = 265$$

12.23 CO_2 and N_2O have different symmetry numbers (σ), which are 2 and 1, respectively.

12.24 The translational partition function is calculated using Equation 12.14 and following steps described in Illustration 12.4.

(a) For a methane molecule trapped in a zeolite of volume

$$\frac{4\pi r^3}{3} = 5.23 \times 10^{-28}\ \mathrm{m^3}:$$

$$q = \frac{(2\pi mkT)^{3/2}\,V}{h^3} =$$

$$= \frac{\left(2\pi \times 16 \times 1.660 \times 10^{-27}\ \text{kg} \times 1.381 \times 10^{-23}\ \text{J K}^{-1} \times 298\ \text{K}\right)^{3/2} \times 5.23 \times 10^{-28}\ \text{m}^3}{\left(6.626 \times 10^{-34}\ \text{J s}\right)^3}$$

$$q = 3.23 \times 10^4$$

(b) For a methane molecule trapped in a flask of volume 100 cm^3 (1×10^{-4} m^3):

$$q = \frac{(2\pi mkT)^{3/2}\,V}{h^3} =$$

$$= \frac{\left(2\pi \times 16 \times 1.660 \times 10^{-27}\ \text{kg} \times 1.381 \times 10^{-23}\ \text{J K}^{-1} \times 298\ \text{K}\right)^{3/2} \times 1 \times 10^{-4}\ \text{m}^3}{\left(6.626 \times 10^{-34}\ \text{J s}\right)^3}$$

$$q = 6.2 \times 10^{27}$$

12.25

$$U - U(0) = -\frac{N}{q}\left(\frac{\partial q}{\partial \beta}\right) \qquad q = 1 + 5\text{e}^{-\beta\varepsilon} + 3\text{e}^{-3\beta\varepsilon}$$

$$U - U(0) = -\frac{N}{1 + 5\text{e}^{-\beta\varepsilon} + 3\text{e}^{-3\beta\varepsilon}}\left(5(-\varepsilon)\text{e}^{-\beta\varepsilon} + 3(-\varepsilon)\text{e}^{-3\beta\varepsilon}\right) = N\varepsilon\left(\frac{5\text{e}^{-\beta\varepsilon} + 3\text{e}^{-3\beta\varepsilon}}{1 + 5\text{e}^{-\beta\varepsilon} + 3\text{e}^{-3\beta\varepsilon}}\right)$$

12.26 The internal energy at a temperature T is given by Equation 12.20 with the term E given by Equation 12.19. For a diatomic molecule such as O_2 or N_2, we must analyze the contributions to the internal energy by translational, rotational and vibrational motions.

Thus: $E_{tot} = E_{trans} + E_{rot} + E_{vib}$ (where E_{elec} has been neglected)

For 1 mol of a gas $kN_A = R$

Therefore, Equation 12.19 is written as:

$$E = \frac{RT^2}{q}\frac{dq}{dT}$$

(i) Let us analyze translational motion:

Substitute Equation 12.14 for q and $\frac{dq}{dT}$ into Equation 12.19 to get for 1 mole of gas:

$$\frac{dq}{dT} = \frac{d}{dT}\left(\frac{(2\pi mkT)^{3/2}V}{h^3}\right) = \left(\frac{(2\pi mk)^{3/2}V}{h^3}\right)\left(\frac{3}{2}T^{1/2}\right)$$

$$E = \frac{RT^2}{\left(\frac{(2\pi mkT)^{3/2}V}{h^3}\right)}\left(\frac{(2\pi mk)^{3/2}V}{h^3}\right)\left(\frac{3}{2}T^{1/2}\right) = \frac{3}{2}RT$$

Therefore, using Equation 12.20, (U –U (0)) $_{\text{trans}}$ = 3/2RT
(ii) Let us analyze rotational motion:

Substitute Equation 12.15a for q and $\frac{dq}{dT}$ into Equation 12.19 to get for 1 mole of gas:

$$\frac{dq}{dT} = \frac{d}{dT}\left(\frac{kT}{\sigma hB}\right) = \left(\frac{k}{\sigma hB}\right)$$

$$E = \frac{RT^2}{\left(\frac{kT}{\sigma hB}\right)}\left(\frac{k}{\sigma hB}\right) = RT$$

Therefore, using Equation 12.20, (U –U (0)) $_{\text{rot}}$ = RT
(ii) Finally, let us analyze vibrational motion:

Substitute Equation 12.13b for q and $\frac{dq}{dT}$ into Equation 12.19 to get for 1 mole of gas:

$$\frac{dq}{dT} = \frac{d}{dT}\left(\frac{kT}{hv}\right) = \left(\frac{k}{hv}\right)$$

$$E = \frac{RT^2}{\left(\frac{kT}{hv}\right)}\left(\frac{k}{hv}\right) = RT$$

Therefore using equation 12.20, (U –U (0)) $_{\text{vib}}$ = RT
In summary, the molar internal energy of a diatomic gas is:
(U –U (0)) = 3/2RT + RT + RT = 7/2 RT

12.27

$$U - U(0) = -\frac{N}{q}\left(\frac{\partial q}{\partial \beta}\right) = N\left(\frac{2\varepsilon e^{-\beta\varepsilon}}{2+2e^{-\beta\varepsilon}}\right)$$

With ε = 121 cm^{-1}~2.40x10^{-21} J, $\beta\varepsilon$=0.580 at T=300 K.
Setting $N = N_A = 6.02\times10^{-23}$mol^{-1} gives 519 J/mol.

12.28 (a) Using the partition function expression from Example 12.1, Equation 12.19 to find the total energy and following steps described in Derivation 12.3 we get:

$$E = \frac{NkT^2}{q}\left(\frac{dq}{dT}\right)$$

$$\left(\frac{dq}{dT}\right) = \mathrm{e}^{-\varepsilon/kT} \times \frac{d}{dT}(-\varepsilon/kT)$$

$$\left(\frac{dq}{dT}\right) = \frac{\varepsilon}{kT^2}\mathrm{e}^{-\varepsilon/kT}$$

Substitution of the expression for q and $\mathrm{d}q/\mathrm{d}T$ into Equation 12.19 we get:

$$E = \frac{NkT^2}{1+\mathrm{e}^{-\varepsilon/kT}}\left(\frac{\varepsilon}{kT^2}\mathrm{e}^{-\varepsilon/kT}\right)$$

$$E = \frac{N\varepsilon\mathrm{e}^{-\varepsilon/kT}}{1+\mathrm{e}^{-\varepsilon/kT}}$$

(b) To find the expression for the constant-volume molar heat capacity we need to find the derivative of the last expression with respect to Temperature according to Equation 12.21

$$\frac{\mathrm{d}E}{\mathrm{d}T} = \frac{\mathrm{d}}{\mathrm{d}T}\left(\frac{N\varepsilon\mathrm{e}^{-\varepsilon/kT}}{1+\mathrm{e}^{-\varepsilon/kT}}\right)$$

If the definition of the molar energy separation is considered the last expression changes to:

$$\frac{\mathrm{d}E}{\mathrm{d}T} = \frac{\mathrm{d}}{\mathrm{d}T}\left(\frac{\varepsilon_m\mathrm{e}^{-\varepsilon_m/RT}}{1+\mathrm{e}^{-\varepsilon_m/RT}}\right)$$

Therefore the derivative $\mathrm{d}E/\mathrm{d}T$ is:

$$\frac{\mathrm{d}E}{\mathrm{d}T} = \frac{1}{\left(1+\mathrm{e}^{-\varepsilon_m/RT}\right)^2}\left[\left(\left(1+\mathrm{e}^{-\varepsilon_m/RT}\right)\frac{\varepsilon_m^2}{RT^2}\mathrm{e}^{-\varepsilon_m/RT}\right) - \left(\frac{\varepsilon_m^2}{RT^2}\mathrm{e}^{-\varepsilon_m/RT}\mathrm{e}^{-\varepsilon_m/RT}\right)\right]$$

$$C_{V,m} = \frac{\mathrm{d}E}{\mathrm{d}T} = \frac{\frac{\varepsilon_m^2}{RT^2}\mathrm{e}^{-\varepsilon_m/RT}}{\left(1+\mathrm{e}^{-\varepsilon_m/RT}\right)^2}$$

Or,

$$C_{V,m} = \frac{\mathrm{d}E}{\mathrm{d}T} = \frac{R\left(\varepsilon_m/RT\right)^2\mathrm{e}^{-\varepsilon_m/RT}}{\left(1+\mathrm{e}^{-\varepsilon_m/RT}\right)^2}$$

(c) A plot of the constant-volume molar heat capacity with temperature is shown below:

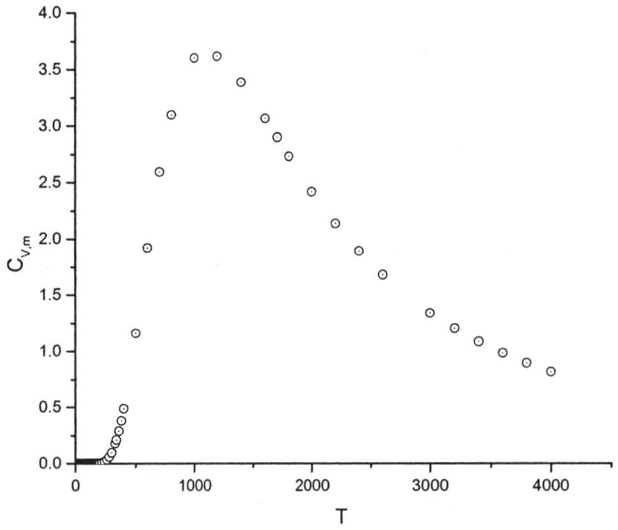

(d) The function has a maximum at around 1100K. If we set up the first derivative of the constant-volume molar heat capacity with temperature to zero we could find the temperature at which the first derivative vanishes.
The evaluation was done by Mathcad 2001 and shown below:

$$\frac{dC_{V,m}}{dT} = \frac{-\varepsilon_m^2 e^{-\varepsilon_m/RT}\left(2RT + 2RTe^{-\varepsilon_m/RT} - \varepsilon_m + \varepsilon_m e^{-\varepsilon_m/RT}\right)}{\left(1 + e^{-\varepsilon_m/RT}\right)^3 R^2 T^4}$$

The temperature at which the first derivative vanishes is $T = 0.4167 \times \dfrac{\varepsilon_m}{R}$.

Using 22 kJ for ε_m, T =1102 K as observed in the graph above.

12.29

$$U_{\text{vib}} = nR\Theta\left(\frac{1}{e^{\Theta/T} - 1}\right) \text{ and } C_V = \left(\frac{\partial U}{\partial T}\right)_V = nR\Theta\left[\frac{-\left(e^{\Theta/T} - 1\right)^{-2}\left(-\Theta e^{\Theta/T}\right)}{T^2}\right]$$

$$= nR\left(\frac{\Theta}{T}\right)^2 \frac{e^{\Theta/T}}{\left(e^{\Theta/T} - 1\right)^2}$$

(b) As can be seen from the graph on the right, C_V approaches its classical value when $T \geq 1.5\ \Theta$

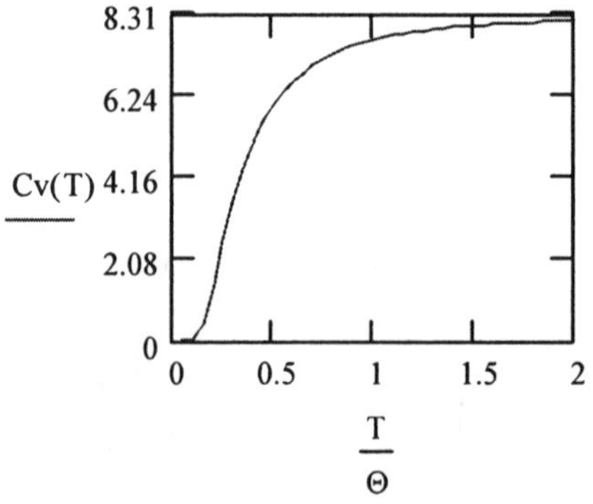

12.30 From Equation 12.29 we can express the change in entropy per micelle as:

$$\Delta S = 100k\ln\frac{V_{solution}}{V_{micelle}}$$

Since volumes are not given, we will have to estimate them. Assuming spherical micelles made of hydrocarbon chains of 12 carbon atoms, the volume of the micelle is:

$$V = \frac{4}{3}\pi r^3.$$

The radius is calculated by assuming a length of 154 pm per C-C bond (Table 9.2).

Therefore:

$$V = \frac{4}{3}\pi r^3 = \frac{4}{3}\pi \times \left(1.85 \times 10^{-9}\ \text{m}\right)^3$$

$$V = 2.64 \times 10^{-26}\ \text{m}^3$$

Assuming 1L solution the change in molar entropy is estimated as:

$$\Delta S = 100k\ln\frac{10^{-3}}{2.64 \times 10^{-26}}$$

$$\Delta S = 7.17 \times 10^{-20}\ \text{J K}^{-1} \times \text{N}_\text{A}$$

$$\Delta S = 43\ \text{kJ K}^{-1}\ \text{mol}^{-1}$$

12.31 The residual molar entropy is given by

$S_m = R\ln W$. In this case $W = 2^{2N}\left(\frac{6}{16}\right)^N = \left(\frac{3}{2}\right)^N \quad \therefore S_m = R\ln\left(\frac{3}{2}\right)^N$

12.32 We write the overall partition function as the product of the translational and rotational partition functions and substitute it into Equation 12.30a. The molar internal energy needed for Equation 12.30a was calculated in Exercise 12.26.

$$q = q^R q^T = \frac{(4\pi IkT)}{\sigma h\hbar} \times \frac{(2\pi mkT)^{3/2} V}{h^3}$$

$$q = \frac{8\pi^2 IkT}{\sigma h^5} \times (2\pi mkT)^{3/2} V$$

We are going to use Equations 12.15b to calculate the moment of inertia of nitrogen:

$$\mu = \frac{m_A m_B}{m_A + m_B} = \frac{m}{2} = \frac{14.01 \times 1.660 \times 10^{-27}\ \text{kg}}{2}$$

$$\mu = 1.163 \times 10^{-26}\ \text{kg}$$

$$I = \mu R^2 = 1.163 \times 10^{-26}\ \text{kg} \times \left(1.10 \times 10^{-10}\ \text{m}\right)^2$$

$$I = 1.41 \times 10^{-46}\ \text{kg m}^2$$

Calculate the overall partition function gives:

$$q = \frac{8\pi^2 IkT}{\sigma h^5} \times (2\pi mkT)^{3/2} V$$

$$q = \frac{8\pi^2 \times 1.41\times 10^{-46}\ \text{kg m}^2 \times 1.381 \times 10^{-23}\ \text{J K}^{-1} \times 298\ \text{K}}{2 \times \left(6.626 \times 10^{-34}\ \text{J s}\right)^5 \times N_\text{A}} \times$$

$$\times\left(2\pi \times 28 \times 1.660 \times 10^{-27}\ \text{kg} \times 1.381 \times 10^{-23}\ \text{J K}^{-1} \times 298\ \text{K}\right)^{3/2} \times 2.44 \times 10^{-2}\ \text{m}^3$$

$$q = 3.02 \times 10^8$$

The volume has been calculated assuming ideal behavior and 1 bar.
Therefore, the molar entropy of nitrogen at 298 K is:

$$S_m = R\ln q + \frac{E}{T}$$

$$S_m = R\ln q + 5/2\text{R}$$

$$S_m = 183\ \text{J K}^{-1}\ \text{mol}^{-1}$$

12.33

$$\beta = \frac{1}{\varepsilon}\ln\left(1+\frac{1}{a}\right).\ \text{For} <\varepsilon> = a\varepsilon,\ \beta = \frac{1}{\varepsilon}\ln 2 = \frac{1}{kT}.$$

When $\varepsilon = 50\,\text{cm}^{-1} = hc\tilde{\nu},\ \ T = \dfrac{hc\tilde{\nu}}{k\ln 2} = 104\,K$.

12.34 From Equation 12.33, the standard molar Gibbs energy can be determined. In Exercise 12.32, we calculated q_m/N_A, so we get:

$$G_m^\varnothing - G_m^\varnothing(0) = -RT\ln\frac{q_\text{m}}{N_\text{A}}$$

$$G_m^\varnothing - G_m^\varnothing(0) = -48.4\ \text{kJ mol}^{-1}$$

12.35 For the distribution given in Equation 12.42, and with q defined as in Equation 12.41, $s = 0.8$ indicates that the chains are mostly helical. However, moving to $s = 1.0$ and larger gives either negative numbers or a singularity. One can also glean this by looking at the two equations cited.

12.36 From Equation 12.41, the partition function is calculated using s values equal to 0.82, 1 and 1.5. From Equation 12.42, values of p_n are found using σ equal to 5E-3 and 5E-2 and the corresponding s values. Once p_n are calculated, the average of n is found at two σ values. A graph of np_n against n is shown below. From the area under the curve we found average values of n of 10.26 and 15.81 for σ equal to 5E-3 and 5E-2 respectively.

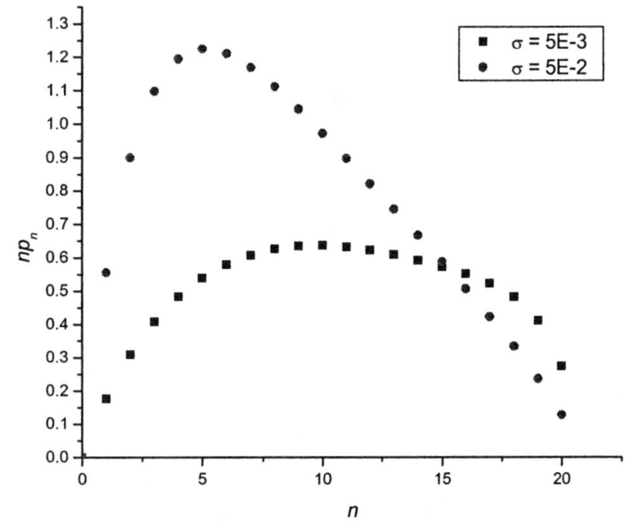

12.37

$$\Theta = \frac{1}{2}\left(1 + \frac{s - 1 + 2\sigma}{\left((s-1)^2 + 4s\sigma\right)^{1/2}}\right) = \frac{1}{2}\left(1 + \frac{1.14 - 1 + 2 \times 0.0033}{\left((1.14-1)^2 + 4 \times 1.14 \times 0.0033\right)^{1/2}}\right) = 0.79$$

12.38 (a) Derivation 12.17 shows the steps involved to deduce the expression for the root-mean square separation of the ends.
(b) The mean separation of the end is calculated by solving the following integral (see Appendix 2):

$$\langle R \rangle = \int_0^\infty R f dR = 4\pi\left(\frac{a}{\pi^{1/2}}\right)^3 \int_0^\infty R^3 e^{-a^2R^2} dR$$

$$\langle R \rangle = 4\pi\left(\frac{a}{\pi^{1/2}}\right)^3 \times \left(\frac{1}{2a^4}\right) = \frac{2}{a\pi^{1/2}}$$

By substituting for a we get:

$$\langle R \rangle = \frac{2}{a\pi^{1/2}} = \left(\frac{8N}{3\pi}\right)^{1/2} l$$

(c) To find their most probable separation of the ends we set the following derivative to zero:

$$\frac{df}{dR} = 0$$

$$\frac{df}{dR} = 4\pi\left(\frac{a}{\pi^{1/2}}\right)^3 \left(2R - 2aR^3\right)e^{-a^2R^2} = 0$$

Now we solve for R:

$$\left(2R - 2aR^3\right) = 0$$

$$R = \frac{1}{a} = \sqrt{\frac{2Nl^2}{3}}$$

The three quantities with N = 4000 and l =154 pm are:
(a) 9.74 nm
(b) 8.97 nm
(c) 7.95 nm

12.39 From Exercise 11 we have:

$$\ln P = \ln\left(\frac{2}{\pi N}\right)^{\frac{1}{2}} - \frac{1}{2}(N+n+1))\ln\left(1+\frac{n}{N}\right) - \frac{1}{2}(N-n+1)\ln\left(1-\frac{n}{N}\right)$$

If we now make the approximation that for small net distances (i.e. $n \ll N$), we can use,

$\ln\left(1\pm\frac{n}{N}\right) \approx \pm\frac{n}{N} - \frac{1}{2}\left(\frac{n}{N}\right)^2$. Substitution into the above equation results in

$\ln P = \ln\left(\frac{2}{\pi N}\right)^{\frac{1}{2}} - \frac{n^2}{2N}$, taking antilogs gives $P \approx \ln\left(\frac{2}{\pi N}\right)^{\frac{1}{2}} \mathrm{e}^{-\left(\frac{n^2}{2N}\right)}$.

Chapter 13: Optical Spectroscopy and Photobiology

13.8 (a) From Equation 13.2a, the frequency is calculated:

$$\nu = \frac{c}{\lambda} = \frac{2.998 \times 10^{8}\ \text{m s}^{-1}}{670 \times 10^{9}\ \text{m}} = 4.74 \times 10^{14}\ \text{s}^{-1}$$

(b) The wavelength of 670 nm expressed in terms of wavenumber is the inverse of 670 nm, or $1.49 \times 10^{6}\ \text{m}^{-1}$ or $1.49 \times 10^{4}\ \text{cm}^{-1}$.

13.9 (a)

$$\nu = \tilde{\nu} c = \frac{92.0 \times 10^{6} s^{-1}}{2.998 \times 10^{8} \text{m}\ s^{-1}} = 0.307\,\text{m}^{-1} = 0.00307\,\text{cm}^{-1}$$

(b) $\lambda = \dfrac{1}{\nu} = 3.26\ \text{m}$

13.10 Following the steps described in Example 13.1, information is provided to calculate the molar absorption coefficient of the coloring matter as shown below:

$$\varepsilon = -\frac{\log T}{[\text{J}]l} = -\frac{\log 0.715}{0.433 \times 10^{-3}\ \text{mol L}^{-1} \times 0.25\ \text{cm}}$$

$$\varepsilon = 1.34 \times 10^{3}\ \text{mol}^{-1}\ \text{L cm}^{-1}$$

$$\varepsilon = 1.34 \times 10^{3}\ \text{mol}^{-1} \times 10^{3}\ \text{cm}^{3}\ \text{cm}^{-1}$$

$$\varepsilon = 1.34 \times 10^{6}\ \text{mol}^{-1}\ \text{cm}^{2}$$

1 L equal 1000 mL, or 1000 cm^3, was used in the conversion above.

13.11 (a)

$$\varepsilon = \frac{A}{bc} = \frac{1.011}{(1.00\ \text{cm})\left(\dfrac{30.2 \times 10^{-3}\text{g}}{(602\ \text{g}\ mol^{-1}) \times (500\ \text{cm}^{3})}\right)} = 1.01 \times 10^{7}\,\frac{\text{cm}^{2}}{\text{mol}}$$

(b)

$$A = \left(1.01\times10^{7}\,\frac{\text{cm}^2}{\text{mol}}\right)\times(1.00\text{ cm})\times 2(1.00\times10^{-7}\,\frac{\text{mol}}{\text{cm}}) = 2.01 \quad \%T = 100\times10^{-A} = 0.96\%$$

13.12 This exercise can be solved using the formula from Exercise 13.10. In this case the thickness has to be determined for two values of transmittance.
The concentration of seawater in mol L^{-1} is calculated, assuming a density of 1 kg L^{-1} at 25°C.
Therefore,

$$[H_2O] = \frac{1000\text{ g L}^{-1}}{18.0\text{ g mol}^{-1}} = 55.5\text{ mol L}^{-1}$$

Solving for l we have:

$$l = -\frac{\log T}{[H_2O]\varepsilon} = -\frac{\log T}{55.5\text{ mol L}^{-1} \times 6.2 \times 10^{-5}\text{ mol}^{-1}\text{ L cm}^{-1}}$$

(a) If the surface light is decreased by 50%, then T is 0.5 and the depth is:

$$l = -\frac{\log T}{[H_2O]\varepsilon} = -\frac{\log 0.5}{55.5\text{ mol L}^{-1} \times 6.2 \times 10^{-5}\text{ mol}^{-1}\text{ L cm}^{-1}}$$

$l = 87$ cm

(b) If the surface light is decreased by 90%, then T is 0.1 and the depth is:

$$l = -\frac{\log T}{[H_2O]\varepsilon} = -\frac{\log 0.1}{55.5\text{ mol L}^{-1} \times 6.2 \times 10^{-5}\text{ mol}^{-1}\text{ L cm}^{-1}}$$

$l = 290$ cm

13.13

$$A = \varepsilon_1 b_1 c_1 = \varepsilon_2 b_2 c_2; \quad \therefore c_2 = \frac{b_1 c_1}{b_2} = \frac{(1.55\text{ cm})\times(25\ \mu\text{g L}^{-1})}{1.18\text{ cm}} = 33\ \mu\text{g L}^{-1}.$$

13.14 (a) The function to consider is then: $\varepsilon = \varepsilon_{\max} e^{-x^2/c}$, in which x is the wavenumber and c is a constant. The half-width at half-height, $x_{1/2}$ is:

$$\frac{1}{2}\varepsilon_{max} = \varepsilon_{max} e^{-x_{1/2}^2/c}$$

$$\ln 2 = \frac{x_{1/2}^2}{c}$$

$$x_{1/2}^2 = c\ln 2$$

$$x_{1/2} = \sqrt{c\ln 2}$$

Therefore, the width at half-height, $\Delta x_{1/2}$ is: $\Delta x_{1/2} = 2\sqrt{c\ln 2}$

Now we need to integrate the function:

$$A = \int \varepsilon \mathrm{d}\tilde{\nu} = \int_{-\infty}^{\infty} \varepsilon_{\max} e^{-x^2/c} \mathrm{d}\tilde{\nu} = \varepsilon_{\max} (c\pi)^{1/2}$$

Since

$$\int_{-\infty}^{\infty} e^{-x^2} \mathrm{d}x = \pi^{1/2}$$

To finish the evaluation of the integral, we need to insert $c = \dfrac{\Delta x_{1/2}^2}{4\ln 2}$ and get:

$$A = \varepsilon_{\max} \left(\frac{\Delta x_{1/2}^2}{4\ln 2} \pi \right)^{1/2}$$

$$A = \varepsilon_{\max} \Delta x_{1/2} \left(\frac{\pi}{4\ln 2} \right)^{1/2}$$

$$A = \varepsilon_{\max} \Delta x_{1/2} 1.0645$$

If we identify $\Delta x_{1/2}$ with $\Delta \tilde{\nu}_{1/2}$, we obtained the desired result.

(b) (i) For a half-width at half-height of 5000 cm^{-1} and ε_{max} of 1×10^4 L mol^{-1} cm^{-1}, we have an integrated absorption coefficient of:

$$A = \varepsilon_{\max} \Delta \tilde{\nu}_{1/2} 1.0645$$

$$A = 1 \times 10^4 \text{ L mol}^{-1} \text{ cm}^{-1} \times 10000 \text{ cm}^{-1} \times 1.0645$$

$$A = 1.06 \times 10^8 \text{ L mol}^{-1} \text{ cm}^{-2}$$

(ii) (i) For a half-width at half-height of 5000 cm^{-1} and ε_{max} of 5×10^2 L mol^{-1} cm^{-1}, we have an integrated absorption coefficient of:

$$A = \varepsilon_{\max} \Delta \tilde{\nu}_{1/2} 1.0645$$

$$A = 5 \times 10^2 \text{ L mol}^{-1} \text{ cm}^{-1} \times 10000 \text{ cm}^{-1} \times 1.0645$$

$$A = 5.32 \times 10^6 \text{ L mol}^{-1} \text{ cm}^{-2}$$

13.15 (a) Assuming ideal gas behavior at STP, $c = P/RT = 0.0446$ mol/L, thus

$$A = \varepsilon bc = (476 \text{ L mol}^{-1}\text{cm}^{-1}) \times (0.030 \text{ cm}) \times (0.0446 \text{ mol L}^{-1}) = 6.4$$

at 100 DU, $A = 2.1$.

(b) Fitting the data (ε in L mol^{-1} cm^{-1} vs. ν in cm^{-1}) gives:

$\varepsilon = 2.9 \times 10^{-16} \exp(0.000126\ \nu)$

Integrating this from 31250 to 34247 cm^{-1} gives integrated ε = 1.43×10^6 mol^{-1} cm^{-2}

13.16 We work through Derivation 13.1 and using the given expression for concentration. We will start with the following integrals:

$$\int_{I_0}^{I}\frac{dI}{I} = -\kappa\int_0^l [J]dx$$

And substitute [J] by $[J] = [J]_0 e^{-x/\lambda}$

$$\int_{I_0}^{I}\frac{dI}{I} = -\kappa\int_0^l [J]_0 e^{-x/\lambda}dx$$

$$\ln\frac{I}{I_0} = -\kappa[J]_0\left[-\lambda e^{-l/\lambda} + \lambda e^{-0/\lambda}\right]$$

Since $l \gg \lambda$ and $e^{-0/\lambda} = 1$, the above expression changes to:

$$\ln\frac{I}{I_0} = -\kappa[J]_0\lambda$$

Or,

$$\log\frac{I}{I_0} = -\varepsilon[J]_0\lambda$$

13.17

Using the given identity for sin(x)sin(y),

$$\mu_{mn} = -e\times\left(\frac{2}{a}\int_0^a x\sin\left(\frac{m\pi x}{a}\right)\sin\left(\frac{n\pi x}{a}\right)dx\right)$$

$$= -\frac{e}{a}\int_0^a\left[x\cos\left(\frac{(m-n)\pi x}{a}\right) - x\cos\left(\frac{(m+n)\pi x}{a}\right)\right]dx$$

Integration yields

$$= -\frac{ea}{\pi^2}\left(\frac{\cos[(m-n)\pi]-1}{(m-n)^2} - \frac{\cos[(m+n)\pi]-1}{(m+n)^2}\right)$$

The relationship is zero when $m+n$ and $m-n$ are even, and non-zero when $m+n$ and $m-n$ are odd. Thus, for n=1 and m=2, $m+n$ and $m-n$ are odd, and $\mu_{mn} \neq 0$. For n=1 and m=3, $m+n$ and $m-n$ are even, and μ_{mn}=0.

13.18 From Equation 13.6b, we can solve for the lifetime of a state as shown below:

(a) $\tau/\text{ps} = \dfrac{5.3\ \text{cm}^{-1}}{0.1\ \text{cm}^{-1}} = 53\ \text{ps}$

(b) $\tau/\text{ps} = \dfrac{5.3\ \text{cm}^{-1}}{1\ \text{cm}^{-1}} = 5.3\ \text{ps}$

(c) $\tau/\text{ps} = \dfrac{5.3\ \text{cm}^{-1}}{1.0\times10^{9}\ \text{s}^{-1}/2.998\times10^{10}\ \text{cm s}^{-1}} = 1.6\times10^{2}\ \text{ps}$

13.19

$$\delta E\delta t = \hbar \text{ or } h\delta\tilde{\nu}c = \frac{\hbar}{\delta t}. \quad \therefore \delta\tilde{\nu}(\text{cm}^{-1}) = \frac{5.3(\text{cm}^{-1})}{\delta t/(\text{ps})}.$$

(a)

$$\delta t = \frac{1}{z} = 0.1 \text{ ps.} \quad \therefore \delta\tilde{\nu} = \frac{5.3}{0.1} = 53 \text{ cm}^{-1}$$

(b)

$$z = \frac{1\times10^{13}/s}{200} = 5\times10^{10}/\text{s}; \quad \therefore \delta t = 2\times10^{-11} = 20 \text{ ps and } \delta\tilde{\nu} = \frac{5.3}{20} = 0.27 \text{ cm}^{-1}.$$

13.20 Since the force constant of the bond in a carbonyl group is given, we can determine the vibrational frequency of the carbonyl group by using Equation 13.8b.

(a) For ^{12}C==^{16}O, the effective mass and the corresponding vibrational frequency are calculated below:

The effective mass of the molecule is:

$$\mu = \frac{\left(12.011 \times 10^{-3} \text{ kg mol}^{-1}\right) \times \left(15.9994 \times 10^{-3} \text{ kg mol}^{-1}\right)}{\left(15.9994 \times 10^{-3} \text{ kg mol}^{-1} + 12.011 \times 10^{-3} \text{ kg mol}^{-1}\right) \times 6.022367 \times 10^{23} \text{ mol}}$$

$$\mu = 1.13919 \times 10^{-26} \text{ kg}$$

Therefore, the vibrational frequency of ^{12}C==^{16}O is:

$$\nu = \frac{1}{2\pi}\left(\frac{k}{\mu}\right)^{1/2}$$

$$\nu = \frac{1}{2\pi}\left(\frac{908 \text{ kg s}^{-2}}{1.13919 \times 10^{-26} \text{ kg}}\right)^{1/2}$$

$$\nu = 4.49 \times 10^{13} \text{ s}$$

(b) For ^{13}C==^{16}O, the effective mass and the corresponding vibrational frequency are calculated below:

The effective mass of the molecule is:

$$\mu_{^{13}C^{16}O} = \frac{\left(13.0033 \times 10^{-3} \text{ kg mol}^{-1}\right) \times \left(15.9994 \times 10^{-3} \text{ kg mol}^{-1}\right)}{\left(15.9994 \times 10^{-3} \text{ kg mol}^{-1} + 13.0033 \times 10^{-3} \text{ kg mol}^{-1}\right) \times 6.022367 \times 10^{23} \text{ mol}}$$

$$\mu_{^{13}C^{16}O} = 1.1911 \times 10^{-26} \text{ kg}$$

Therefore, the vibrational frequency of ^{13}C==^{16}O is:

$$\nu = \frac{1}{2\pi}\left(\frac{k}{\mu}\right)^{1/2}$$

$$\nu = \frac{1}{2\pi}\left(\frac{908 \text{ kg s}^{-2}}{1.1911 \times 10^{-26} \text{ kg}}\right)^{1/2}$$

$$\nu = 4.39 \times 10^{13} \text{ s}$$

13.21

$$\nu = \frac{1}{2\pi}\sqrt{\frac{k}{\mu}} \quad \text{or} \quad \tilde{\nu} = \frac{1}{2\pi c}\sqrt{\frac{k}{\mu}} \quad \therefore k = 4\pi^2\tilde{\nu}^2c^2\mu \text{ for HF we have,}$$

$$k_{\mathrm{HF}} = 4\pi^2(4141.3\ \mathrm{cm}^{-1})^2(2.998\times10^{10}\,\mathrm{cm/s})^2\left(\frac{1.008\times19.00}{1.008+19.00}\right)\times1.661\times10^{-27}\,\mathrm{kg}$$

$$= 968\ \mathrm{kg\ s}^{-2} = 968\ \mathrm{Nm}^{-1}$$

$$\tilde{\nu}_{DF} = \tilde{\nu}_{HF}\sqrt{\frac{\mu_{HF}}{\mu_{DF}}} = 4141.3\ \mathrm{cm}^{-1}\sqrt{\frac{1.008\times19.00(2.016+19.00)}{(1.008+19.00)(2.016\times19.00)}} = 3001\ \mathrm{cm}^{-1}$$

The others are done completely analogously and are presented in the table below.

	HF	HCl	HBr	HI
$\tilde{\nu}/\mathrm{cm}^{-1}$	4141.3	2988.9	2649.7	2309.5
$k/(\mathrm{N\ m}^{-1})$	968	516	412	314
$\tilde{\nu}_{\text{D-halide}}/\mathrm{cm}^{-1}$	3001	2142	1885	1639

13.22 Section 13.4 mentions that homonuclear diatomic molecules are infrared inactive. From the given list, only hydrogen and nitrogen are infrared inactive.

13.23 Number of normal vibrations = $3N$–5 (linear) or $3N$–6 (nonlinear)
(a) NO_2 (bent): 3 (b) N_2O (linear): 4 (c) C_6H_{12}: 48 (d) C_6H_{14}: 54

13.24 The uniform expansion of the benzene ring can be described by Figure 13.20. It seems that the exclusion rule applies to benzene because it is centrosymmetric. Then, benzene is infrared inactive and Raman active.

13.25 (a) If linear, then 7 normal modes.
(b) Based on the data, it would have to be the one with a center of symmetry which would exclude mutual absorptions in the IR and Raman.

13.26 Compare the OH stretch band by Raman spectroscopy before and after hydrolysis of cellulose to determine the content of cellulose.

13.27 The weak absorption at 30,000 cm^{-1} is consistent with an $n \rightarrow \pi^*$ transition due to the lone pairs of oxygen while the strong absorption can be assigned to a $\pi \rightarrow \pi^*$ transition arising from the pi orbitals.

13.28 Glycine and Cysteine are the only amino acids in the figure lacking a benzene ring.

13.29 (a) Lengthening the chain increases the number of pi orbitals and decreases the spacing between orbitals, which lowers the energy of the first absorption band.
(b) Red Shift – to lower energy.

13.30 The peak power according to the definition is:

$$P_{peak} = \frac{0.10\ \text{J}}{3.0 \times 10^{-9}\ \text{s}} = 3.3 \times 10^{7}\ \text{W}$$

The average power output is then:

$$P_{average} = 0.1\ \text{J} \times 10\ \text{s}^{-1} = 1.0\ \text{W}$$

13.31 (a) We need the ratio $\frac{I_{rod}}{I_{cc}}$ for DNA. The structure factor P depends on both conformation and scattering angle. Thus we have,

$$P \approx 1 - \frac{16\pi^2 R_g^2 \sin^2(\theta/2)}{3\lambda^2} = \frac{3\lambda^2 - 16\pi^2 R_g^2 \sin^2(\theta/2)}{3\lambda^2}$$

However, from Exercise 11.47,

R_g for a rod of length l is $\frac{l}{\sqrt{12}}$ and is $\frac{l}{2\pi}$ for a closed circle.

This is true because the circumference of the circle has to be the same as the length l. Therefore,

$$\frac{I_{\text{rod}}}{I_{\text{cc}}} = \frac{P_{\text{rod}}}{P_{\text{cc}}} = \frac{3\lambda^2 - \frac{4}{3}\pi^2 l^2 \sin^2(\theta/2)}{3\lambda^2 - 4l^2 \sin^2(\theta/2)},$$ plugging in $\lambda = 488$ nm and $l = 250$ nm

$\theta/°$	20	45	90
$I_{\text{rod}}/I_{\text{cc}}$	0.976	0.876	0.514

(b) The experiment should be conducted at an angle of 90°, which gives the smallest ratio as well as the largest difference from the other two.

13.32 (a) A graph of I_f/I_0 against time is shown below. The fitting of the curve yields an observed lifetime of 6.52 ns.

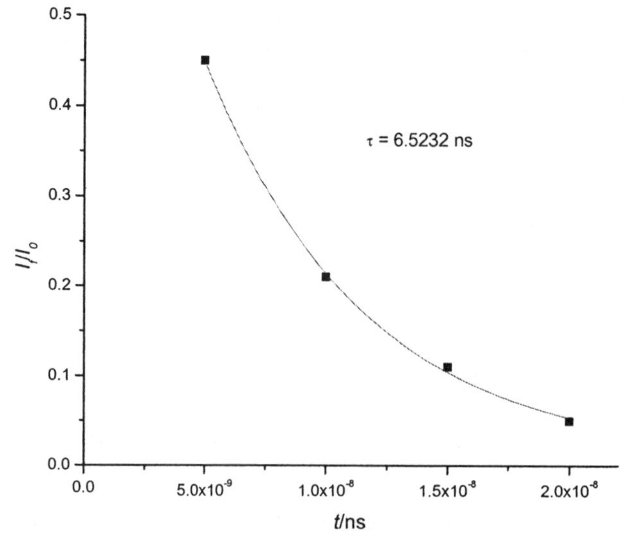

(b) From Equation 13.22, we can determine the fluorescence rate constant:

$$k_f = \frac{\Phi_f}{\tau_0} = \frac{0.70}{6.52 \times 10^{-9}\ \text{s}} = 1.07 \times 10^8\ \text{s}^{-1}$$

13.33 Since it's very dilute, we make take the density of solvent equal to the density of the solution. Thus,

$$1\times10^{-15}\,\text{L(solution)}\times\frac{1\ \text{kg}}{\text{L}}\times\frac{0.1\times10^{-3}\,\text{g(impurity)}}{\text{kg(solution)}} = 1\times10^{-19}\,\text{g(impurity)}, \text{ and}$$

$$\frac{1\times10^{-19}\,\text{g(imp)}}{100\ \text{g(imp)/mol}} = (1\times10^{-21}\,\text{mol(imp)})\times(6.022\times10^{23}\,\text{molecule/mol}) = 602$$

molecules of impurity. This would be unsuitable for a single molecule experiment.

13.34 Assuming the fluorescent dye withstands 10^6 excitations by photons, the total energy delivered by 488 nm radiation is:

$E = h/c\lambda =$

$6.626 \times 10^{-34}\ \text{J s} \times 2.998 \times 10^8\ \text{m s}^{-1} / 4.88 \times 10^{-7}\ \text{m} = 4.07 \times 10^{-19}\ \text{J/photon}$

$E = 4.07 \times 10^{-19}\ \text{J/photon} \times 10^6\ \text{photons} = 4.07 \times 10^{-13}\ \text{J}$

Next, calculate the total number of photons:

$$\frac{1.0\times10^{-3}\ J\,s^{-1}}{4.07\times10^{-19}\ J\ photon^{-1}} = 2.5\times10^{15}\ photons\ s^{-1}$$

Therefore, photobleaching will be relevant after 0.40 ns as shown below:

$$t = \frac{10^6\ photons}{2.5\times10^{15}\ photons\ s^{-1}} = 4.0 \times 10^{-10}\ \text{s}$$

13.35

$$k_{fl} = \frac{1}{\tau_{fl}} = 1\times10^9\ /\text{s};\quad k_{\text{ph}} = 1\times10^3\ /\text{s and } k_{\text{uni}} = 1.7\times10^4\ /\text{s}$$

This indicates that $k_{fl} >> k_{\text{uni}}$ and $k_{\text{ph}} \sim k_{\text{uni}}$, it seems likely that the triplet state is the precursor.

13.36 With the power of the source and the time of the experiment, we can determine the total energy delivered to the sample:

$E = 100\ \text{W} \times 45\ \text{min} \times 60\ \text{s min}^{-1} = 2.7 \times 10^5\ \text{J}$

The energy delivered by 490 nm radiation per photon is:

$E = h/c\lambda =$

$6.626 \times 10^{-34}\ \text{J s} \times 2.998 \times 10^8\ \text{m s}^{-1} / 4.90 \times 10^{-7}\ \text{m} = 4.05 \times 10^{-19}\ \text{J/photon}$

Next, calculate the total number of photons:

$$\frac{2.7 \times 10^5\ \text{J}}{4.05 \times 10^{-19}\ \text{J photon}^{-1}} = 6.66 \times 10^{23}\ \text{photons}$$

60% of these photons were absorbed $= 3.99 \times 10^{23}$ photons or 0.66 mol of photons. Then the ratio of 0.344 mol of absorbing substance by 0.66 mol of photons will be the quantum yield: 0.52.

13.37 The initial amount of oxalic acid present $= \dfrac{5.232\ \text{g}}{90.036\ \text{g/mol}} = 5.81\times10^{-2}\text{mol,Ox}$.

Titration gives:

$$(0.212\ \text{M}\times0.017\ \text{L}) = (3.604\times10^{-3}\text{mol, perman.})\times\frac{5\ \text{mol,Ox}}{2\ \text{mol,perman.}} = 9.01\times10^{-3}\text{mol,Ox}$$

The amount of Ox irradiated is 0.0581–0.0091 = 0.0491 mol. Thus the incidence of photons is:

$$(0.0491\ \text{mol Ox})\times\left(\frac{1\ \text{mol photons}}{0.53\ \text{mol Ox}}\right)\times(6.023\times10^{23}\ \text{mol}^{-1})$$

$$= 5.6\times10^{22}\,\frac{\text{photons}}{300\ \text{s}} = 1.9\times10^{20}\text{photons s}^{-1}$$

13.38 A graph of I_{phos} versus [Q] is shown below:

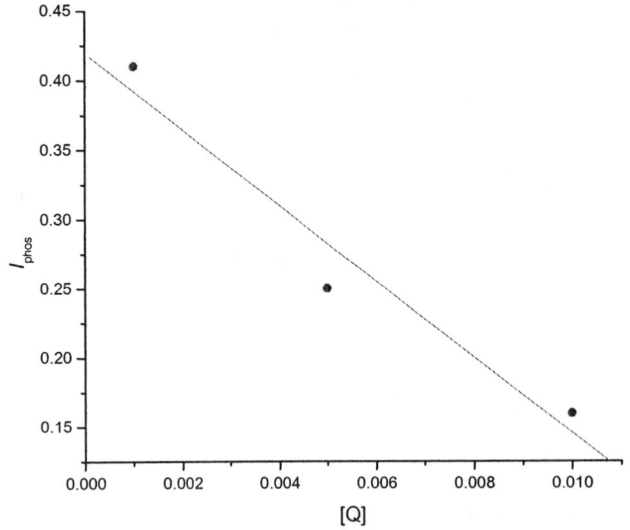

The slope of the graph, as shown in Figure 13.51, is $\tau_0 k_q = 27.37\ \text{mol L}^{-1}$. Therefore, the value of k_q is:

$$k_q = \frac{27.37\ \text{mol L}^{-1}}{29 \times 10^{-6}\ \text{s}} = 9.4 \times 10^5\ \text{mol L}^{-1}\ \text{s}^{-1}$$

13.39

A linear fit of $\frac{1}{\tau}$ vs [Q] gives $k_q = 9.2 \times 10^9\ \text{L mol}^{-1}\text{s}^{-1}$ as the slope and the intercept is $k_{fl} = 3.65 \times 10^6\,\text{s}^{-1}$. Since fluorescence is first order, we have $t_{1/2} = \ln 2 / k_{fl} = 1.9 \times 10^{-7}$.

13.40 If the system can be described by Equation 13.26, then a graph of $1/\varepsilon_T$ versus R^6 should yield a straight line with slope equal to $1/R_0^6$

The results are shown below:

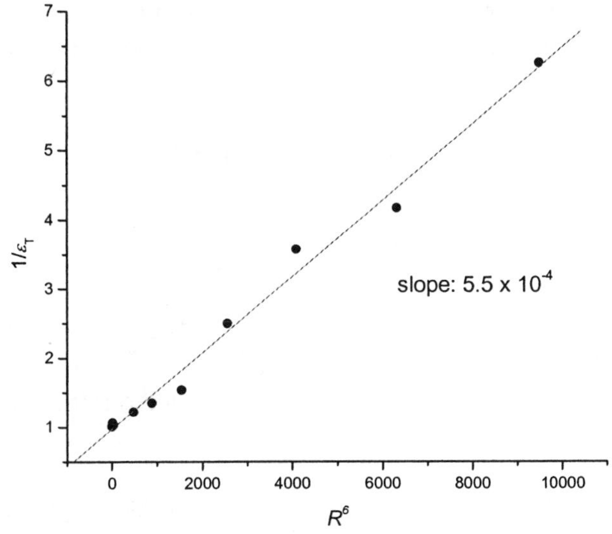

A slope of 5.5×10^{-4} will yield a value of $R_0^6 = 1818$. R is 3.49 nm, which is within the range of the technique. The possible value of R and the linear behavior observed above seems to indicate the data could indeed be described by Equation 13.26.

13.41 Flux = $4 \times 10^3\,\text{mm}^{-2}\text{s}^{-1}$. The reductions proceed as follows: $0.7 \times 0.75 \times 0.91 \times 0.57 \times 4000 = 1089\ \text{mm}^{-2}\text{s}^{-1}$ photons. In 0.1 s over a 40mm^2 area, we have $1089 \times 40 \times 0.1 = 4357$ photons.

13.42 This exercise can be solved by following the steps described in Illustration 13.4. From Equation 13.25, we calculate the efficiency ε:

$$\varepsilon = 1 - \frac{10 \times 10^{-12}\ \text{s}}{1 \times 10^{-9}\ \text{s}}$$

$$\varepsilon = 0.99$$

From Equation 13.26, we solve for R:

$$\varepsilon_T = \frac{R_0^6}{R_0^6 + R^6}$$

$$\frac{1}{\varepsilon_T} = 1 + \frac{R^6}{R_0^6}$$

$$R = \left(\frac{1}{\varepsilon_T} - 1\right)^{1/6} \times R_0 = 2.60\ \text{nm}$$

13.43 Since a reducing agent gives up an electron, its strength must be inversely related to ionization energy. A molecule in an excited electronic state has a lower ionization energy than in the ground state since an electron has been promoted to a higher energy, previously unoccupied orbital. Thus, it is more easily removed. A picture in term of MO's:

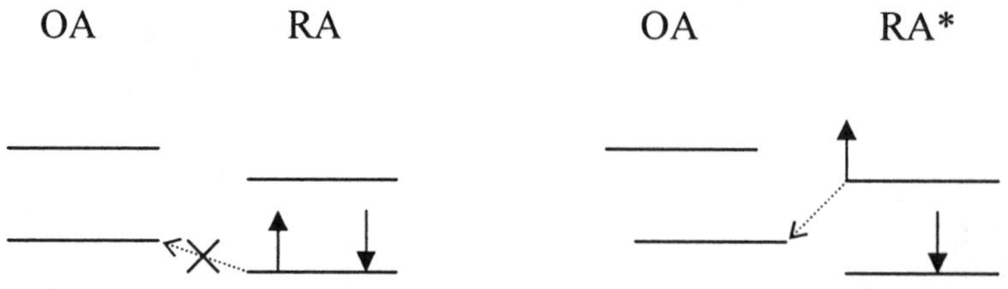

(GS: NO electron transfer) (ES: Spontaneous electron transfer)

13.44 It seems that the new band is related to the formation of singlet oxygen. To confirm, the hypothesis experiments should be carried out by using different oxygen concentrations.

Chapter 14: Magnetic Resonance

14.9 For an electron in a magnetic field, the energy separation of spin states is given by

$E = m_s g B_e \boldsymbol{B}_0$,

where g is the Lande *g-factor* (2.0023193 for a free electron).

$\boldsymbol{B_o}$ is the Bohr Magneton (9.274 × 10^{-24} for a free electron).

$\Delta E = E_\alpha - E_\beta = g\, B_e \boldsymbol{B_o}$

So therefore:

$$\Delta E = (2.0023193)\times(9.274\times10^{-24}\ \text{J/T})\times(0.300\ \text{T})$$

$$= 5.57 \times 10^{-24}\ \text{J}$$

14.10 The nucleus ^{32}S, with a spin of 3/2, will have $2\times\frac{3}{2}+1=4$ orientations. These orientations are determined by m_I. In this case, m_I values are: 3/2, 1/2, −1/2 and −3/2.

The energies of the nuclear spin states can be calculated by using Equation 14.7 as shown below:

$$E = -\left(0.4289 \times 5.051 \times 10^{-27}\ \text{J T}^{-1} \times 7.500\ \text{T}\right) m_I$$

$$E = -1.624 \times 10^{-26}\ \text{J}\ m_I$$

14.11 Given Equations 14.5 and 14.7, we can write:

$E = -\gamma_N\, h\, B_0\, \text{m}_I$

By erforming cancellation and rearranging, we get:

$$\gamma_N = \frac{E}{hB_0 m_I} \sim \frac{\text{J}}{\text{J}\times\text{s}\times\text{T}} = \text{T}^{-1}\text{s}^{-1}$$

Therefore, the unit is 1/(T s) or T Hz and then to restate in SI base units, we use Hz = 1/s, T = kg s^{-2} Å^{-1} and 1 m = 1×10^{10} Å to allow us to restate the unit as 1×10^{10} s m kg^{-1}.

14.12 From Equations 14.5 and 14.7, we can see the following relationship applies, which will be used to solve for the g-value of the nucleus:

$$\gamma_N \hbar = g_I \mu_N$$

$$g_I = \frac{\gamma_N \hbar}{\mu_N} = \frac{1.0840 \times 10^8 \text{ T}^{-1} \text{ s}^{-1} \times 1.05457 \times 10^{-34} \text{ J s}}{5.051 \times 10^{-27} \text{ J T}^{-1}}$$

$$g_I = 2.263$$

14.13 We seek to ratio the difference in population for the alpha and beta states caused by electron spin. The value of $(N_\beta - N_\alpha)/N$ can also be written as $(N_\beta - N_\alpha)/(N_\beta + N_\alpha)$. With $h\nu = g_e\, \mu_b B_0$ we can rearrange and substitute to say $\frac{N_\alpha}{N_\beta} = e^{-gB_0/kT}$ .So, for a 0.300 T field the ratio is 0.998.

The value of $(N_\beta - N_\alpha)/N$ can also be written as $(N_\beta - N_\alpha)/(N_\beta + N_\alpha)$. With $h\nu = g_e\, \mu_b B_0$ we can rearrange and substitute to say $N_u/N_l = e^{-g\, B_0/kt}$. For a 1.1 T field, the ratio is determined by e^{-f} where the factor f is given by

f = (2.0023195)(9.274×10^{-24} J T^{-1})(1.1 T) / (1.38×10^{-23} J K^{-1})(214 K)

In the stronger field, the ratio = 0.999999182.

14.14 The resonance frequency can be calculated from Equation 14.10, as shown below:

$$\nu = \frac{g_e \mu_B B_0}{h} = \frac{2.0023 \times 9.274 \times 10^{-24} \text{ J T}^{-1} \times 0.330 \text{ T}}{6.626 \times 10^{-34} \text{ J s}}$$

$$\nu = 9.248 \times 10^9 \text{ s}^{-1}$$

$$\nu = 9.248 \text{ GHz}$$

See Section 14.12 for a discussion about the g-value.

The corresponding wavelength is calculated below:

$$\lambda = \frac{c}{\nu} = \frac{2.998 \times 10^8 \text{ m s}^{-1}}{9.248 \times 10^9 \text{ s}^{-1}} = 3.241 \times 10^{-2} \text{ m}$$

$\hbar = \frac{h}{2\pi}$

14.15 For ^{1}H, $(N_\alpha - N_\beta)/N = (\gamma_N h\, B_0)/(4\pi k T)$

= [(26.7522128×10^7 T^{-1}s^{-1})(6.626×10^{-34} J s)(10 T)] /[(4π)(1.38066×10^{-23} J K^{-1})(298 K^{-1})]

= 3.42×10^{-5}

and for ^{13}C, in similar fashion,

= [(6.728284×10^{7}T^{-1}s^{-1})(6.626×10^{-34} J s)(10 T)] /[(4π)(1.38066×10^{-23} J K^{-1})(298 K^{-1})]

= 8.585×10^{-6}

14.16 A population difference is proportional to the applied field, as shown by Equation 14.14. By substituting Equation 14.12 into 14.14, we get:

$$N_\alpha - N_\beta \approx \frac{Nh\nu}{2kT}$$

Therefore, the relative population difference in the two spectrometers is given below:

$$\frac{N_\alpha - N_\beta\,(800\text{ MHz})}{N_\alpha - N_\beta\,(60\text{ MHz})} = \frac{800}{60} = 13$$

The nature of the nuclide is irrelevant.

14.17 $\Delta E = (\gamma_N B_o h)/2\pi$

$= 2.5177\times10^{8}\text{ T}^{-1}\text{s}^{-1})(1.05457\times10^{-34}\text{ J s})(8.200\text{ T})$

$= 2.1771745\times10^{-25}\text{J} = h\nu$

Therefore,

$\nu = 3.507\times10^{8}\text{ Hz}$

14.18 Using Equation 14.12 and the relation we derived earlier ($\gamma_N\hbar = g_I\mu_N$), we obtain:

$$\nu = \frac{\gamma_N B_0}{2\pi} = \frac{g_I\mu_N B_0}{2\pi\hbar}$$

Or,

$$\nu = \frac{g_I\mu_N B_0}{h} = \frac{0.4036\times5.051\times10^{-27}\text{ J T}^{-1}\times15.00\text{ T}}{6.626\times10^{-34}\text{ J s}}$$

$\nu = 4.615\times10^{7}\text{ s}^{-1}$

$\nu = 46.15\text{ MHz}$

14.19 $\Delta E = (\gamma_N B_o h)/2\pi = h\nu$, so

$\nu = (\gamma_N B_o)/2\pi$, so

$550.0\times10^{6}\text{ Hz} = (26.7522128\times10^{7}\text{T}^{-1}\text{s}^{-1})(\text{B}_o)/2\pi$

Therefore, $B_o = 12.97$ T

~ 13 T

14.20 Using Equation 14.18 and following the steps described in Illustration 14.1, we obtain:

$$\nu - \nu^0 = \frac{420\times10^{6}\text{ Hz}}{10^{6}}\times6.33 = 2.66\text{ kHz}$$

14.21 The key relationship is: $\nu = \dfrac{\gamma B}{2\pi}$

Chemical shift is the difference between the signal being examined and the reference signal, hence $\delta = \left((\nu - \nu^0)/\nu^0\right) \times 10^6$, where ν^0 is the frequency of the reference standard.

a) Since both ν and ν^0 are directly proportional to applied field (*B*), the ratio δ is independent of the applied field

b) For the difference in terms of frequency $(\nu - \nu^0)$, since (again) both scale directly with *B*, the difference will scale directly with *B*, i.e., by a factor of 800/60 = 13.

So: $(\nu - \nu^0)$ (*B*=800 MHz) = $13 \times (\nu - \nu^0)$ (*B*=60 MHz)

14.22 Section 14.2 discusses the factors that control the intensities of spin-flipping transitions. The intensity of absorption in NMR spectra not only increases with the applied field, but with the frequency of the spectrometer. In an NMR spectrum, the splitting between signals is proportional to the spectrometer frequency. A spectrometer of higher frequency facilitates the analysis of complex overlapping signals.

14.23 $B_{\text{loc}} = (1 - \sigma)\, B_o$

$|\Delta B_{\text{loc}}| = |\delta(\text{CH}_3) - \delta(\text{CHO})| \times B = (7.7 \times 10^6) \times B$

(a) $\Delta B_{\text{loc}} = (7.7 \times 10^6) \times (1.5\ \text{T}) = 1.2 \times 10^7\ \text{T}$

(b) $\Delta B_{\text{loc}} = (7.7 \times 10^6) \times (6.0\ \text{T}) = 4.6 \times 10^7\ \text{T}$

14.24 Using Figure 14.4, we find that the δ values for the methyl and aldehydic protons differ by 8. Therefore, we can determine the splitting between the mentioned protons by using Equation 14.18:

(a) For a spectrometer operating at 300 MHz:

$$\nu - \nu^0 = \frac{300 \times 10^6\ \text{Hz}}{10^6} \times 8 = 2400\ \text{Hz}$$

(b) For a spectrometer operating at 550 MHz:

$$\nu - \nu^0 = \frac{550 \times 10^6\ \text{Hz}}{10^6} \times 8 = 4400\ \text{Hz}$$

14.25 Splitting patterns are given by following Pascal's triangle from the zeroth level (top) down to the level appropriate for the number of protons.

1
1 1
1 2 1
1 3 3 1
1 4 6 4 1
1 5 10 10 5 1
1 6 15 20 15 6 1
1 7 21 35 35 21 7 1 This row reveals the eight-peak multiplet with intensity ratios as at left.

14.26 A 1 2 3 2 1 quintet for the two equivalent nitrogen nucleus, and a 1 3 6 7 6 3 1 septet for the three equivalent nitrogen nucleus. See Example 14.3.

14.27 If one examines the possible spin arrangements:

d u u
u u u
d d u

d
u
u

This *jj* spin coupling results in a triplet with intensity ratio of 1:2:1.

14.28 In Exercise 14.24, we determined the splitting between methyl and aldehydic protons. Since J does not change and it is equal to 2.90 Hz, the ^{1}H-NMR spectrum of ethanal is shown below in a spectrometer operating at 300MHz and 550 MHz.

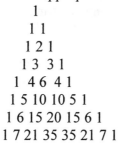

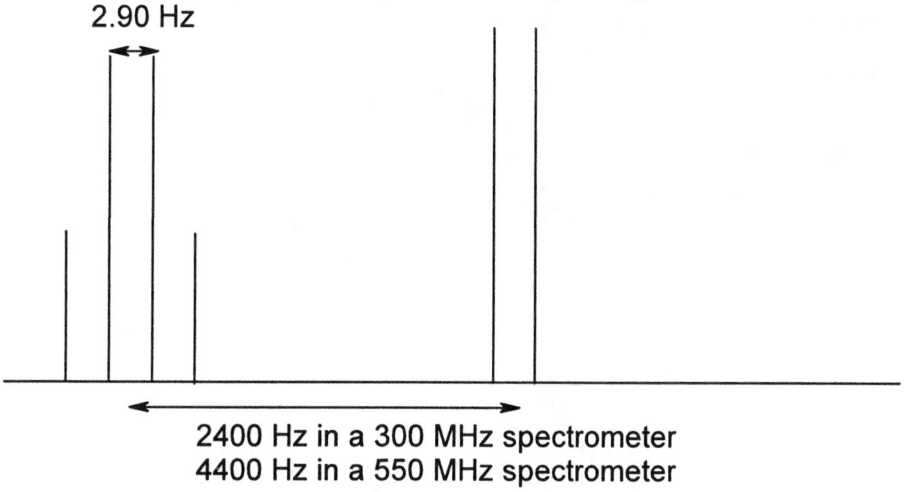

14.29 The A protons are split into a strong triplet by M and weakly toward pentet by X. The result would be a broad triplet at low resolution. At high resolution the fine structure would begin to show the additional splitting of the pentet.
The M protons are strongly split into a quartet by the proximity of A then very weakly into pentet fine structure by X. At low and even good resolution, this peak is likely a straightforward quartet. Very high resolution could reveal the additional splitting. X is split into a weak triplet and a superimposed quartet of reasonable intensity (the resonance is medium in strength but 2 nuclei removed.

14.30 Starting from the Karplus equation, evaluate the first and second derivative with respect to ϕ.

The first derivative of $\frac{d\,^3J_{HH}}{d\phi}$ is $-B\sin\phi - 2C\sin 2\phi$.

Since $\sin 2\phi$ is $= 2\sin\phi\cos\phi$, the first derivative is:

$$\frac{d\,^3J_{HH}}{d\phi} = -B\sin\phi - 4C\sin\phi\cos\phi$$

Then set the result to zero:

$B\sin\phi + 4C\sin\phi\cos\phi = 0$

If the equation passes through a minimum when $\cos\phi$ = B/4C, then ϕ = arcosB/4C and $\sin\phi = \sqrt{1 - B^2/16C^2}$ because $\sin\phi = \sqrt{1 - \cos^2\phi}$

Substitution of $\cos\phi$ and $\sin\phi$ into $B\sin\phi + 4C\sin\phi\cos\phi = 0$ yields:

$$B\sqrt{1 - B^2/16C^2} + 4C\sqrt{1 - B^2/16C^2}\,(B/4C) = 0$$

Clearly, $\cos\phi$ = B/4c satisfies the equation above.
To confirm the extremum is a minimum, evaluate the second derivative:

The second derivative of $\frac{d^2(^3J_{HH})}{d\phi^2} =$ is $-B\cos\phi - 4C\cos 2\phi$.

Since $\cos 2\phi$ is $= 2\cos^2\phi - 1$

Thus, the second derivative is:

$$\frac{d^2(^3J_{HH})}{d\phi^2} = -B\cos\phi - 4C(2\cos^2\phi - 1)$$

Set the result to zero:

$-B\cos\phi - 4C(2\cos^2\phi - 1) = 0$

Substitute $cos\phi$ into the equation above:

$$-B\left(\frac{B}{4C}\right) - 4C\left(2\left(\frac{B}{4C}\right)^2 - 1\right) = 0$$

$$-\frac{B^2}{4C} - 4C\left(\frac{B^2}{8C^2} - 1\right) = 0$$

$$-\frac{B^2}{4C} - \frac{B^2}{2C} + 4C = 0$$

Or,

$$-\frac{B^2}{C}\left(\frac{1}{4} + \frac{1}{2}\right) + 4C = 0$$

$$-\frac{3B^2}{4C} + 4C = 0$$

The above yields a positive result if $4C > 3B^2$.

14.31 $\Delta\nu = 1/(4\pi t)$

From Heisenberg, $Tc = 1.4142136 / \pi\ (550\ \text{MHz})\ (4.8-2.7)$

$= 3.8975 \times 10^{-10}\ \text{s} = 0.4\text{ns}$

14.32 If a plot of $[\text{I}]_0$ versus $\delta\nu^{-1}$ is a straight line with slope $[\text{E}]_0\Delta\nu$ and y-intercept K_I, we need to obtain the following relation:

$$[\text{I}]_0 = -K_I + \frac{[\text{E}]_0\Delta\nu}{\delta\nu}$$

Start with the given expression of resonance frequency:

$$\nu = \left(\frac{[\text{I}]}{[\text{I}] + [\text{EI}]}\right)\nu_I + \left(\frac{[\text{EI}]}{[\text{I}] + [\text{EI}]}\right)\nu_{EI}$$

Substitute $[\text{I}] + [\text{EI}] = [\text{I}]_0$:

$$\nu = \left(\frac{[\text{I}]_0 - [\text{EI}]}{[\text{I}]_0}\right)\nu_I + \left(\frac{[\text{EI}]}{[\text{I}]_0}\right)\nu_{EI}$$

And solve for $[\text{I}]_0$:

$$\nu[\text{I}]_0 = [\text{I}]_0\nu_I - [\text{EI}]\nu_I + [\text{EI}]\nu_{EI}$$

$$[\text{I}]_0(\nu - \nu_I) = [\text{EI}](\nu_{EI} - \nu_I)$$

$$[\text{I}]_0 = \frac{[\text{EI}]\Delta\nu}{\delta\nu}$$

We need to relate $[\text{I}]_0$ with the equilibrium constant and the initial concentration of enzyme.

Then write the equilibrium expression:

$$K_I = \frac{[\mathrm{E}][\mathrm{I}]}{[\mathrm{EI}]} = \frac{([\mathrm{E}]_0 - [\mathrm{EI}])([\mathrm{I}]_0 - [\mathrm{EI}])}{[\mathrm{EI}]}$$

Since the initial concentration of I, $[\mathrm{I}]_0$ is much greater than the initial concentration of E, $[\mathrm{E}]_0$, we can neglect the concentration of [EI] compared to $[\mathrm{I}]_0$.

Therefore, the expression above changes to:

$$K_I = \frac{([\mathrm{E}]_0 - [\mathrm{EI}])[\mathrm{I}]_0}{[\mathrm{EI}]}$$

Substitute $[\mathrm{I}]_0 = \dfrac{[\mathrm{EI}]\Delta\nu}{\delta\nu}$:

$$K_I = \frac{([\mathrm{E}]_0 - [\mathrm{EI}])[\mathrm{I}]_0}{[\mathrm{EI}]}$$

$$K_I = \frac{\left([\mathrm{E}]_0 - [\mathrm{I}]_0\delta\nu\Delta\nu^{-1}\right)[\mathrm{I}]_0}{[\mathrm{I}]_0\delta\nu\Delta\nu^{-1}}$$

Solve for $[\mathrm{I}]_0$:

$$K_I = \frac{[\mathrm{E}]_0}{\delta\nu\Delta\nu^{-1}} - [\mathrm{I}]_0$$

$$[\mathrm{I}]_0 = \frac{[\mathrm{E}]_0}{\delta\nu\Delta\nu^{-1}} - K_I$$

14.33

$$\theta = \omega t = \frac{g_I\mu_N B}{\hbar}t$$

$$B = \frac{\hbar\theta}{g_I\mu_N t} = \left(\frac{(1.0546\times10^{-34}\,\mathrm{J\,s})(\pi/2)}{(5.586)(5.0508\times10^{-27}\,\mathrm{J\,T^{-1}})(10\times10^{-6}\,\mathrm{s}}\right) = 5.9\times10^{-4}\,\mathrm{T}$$

An H nucleus has been assumed above.

14.34 (a) The nuclear overhauser effect is observed because methione has a methyl group next to an S atom which is not present in tryptophan. (b) tryptophan and tyrosine have very similar functional groups then saturation effect is not observed.

14.35 $\nu(\mathrm{Hz}) = (2.5177\times10^8/2\pi\ \mathrm{T^{-1}s^{-1}})[(9.4\ \mathrm{T} + G(\mathrm{T/m})\times(.08\ \mathrm{m})]$

The 100 Hz of separation $= [G(\mathrm{T/m})(.08\ \mathrm{m})(2.5177\times10^8\mathrm{T^{-1}s^{-1}})]/2\pi$

Solve for the gradient G

$G = 31.2\mu\,\mathrm{T/m}$

14.36 See Section 14.8. The MRI image of the disk could be similar to the drawing below.

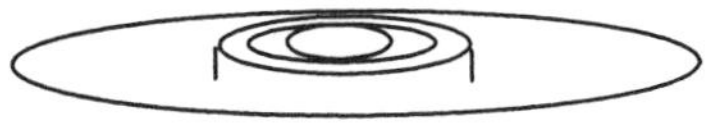

14.37 Off diagonal peaks indicate coupling between H's on various carbons. Thus, the peaks at (4,2) and (2,4) indicate that the H's on the adjacent CH_2 units are coupled. The peaks at (1,2) and (2,1) indicate that H's on the CH_3 and central CH_2 units are coupled.

14.38 A COSY spectrum for alanine is sketched below:

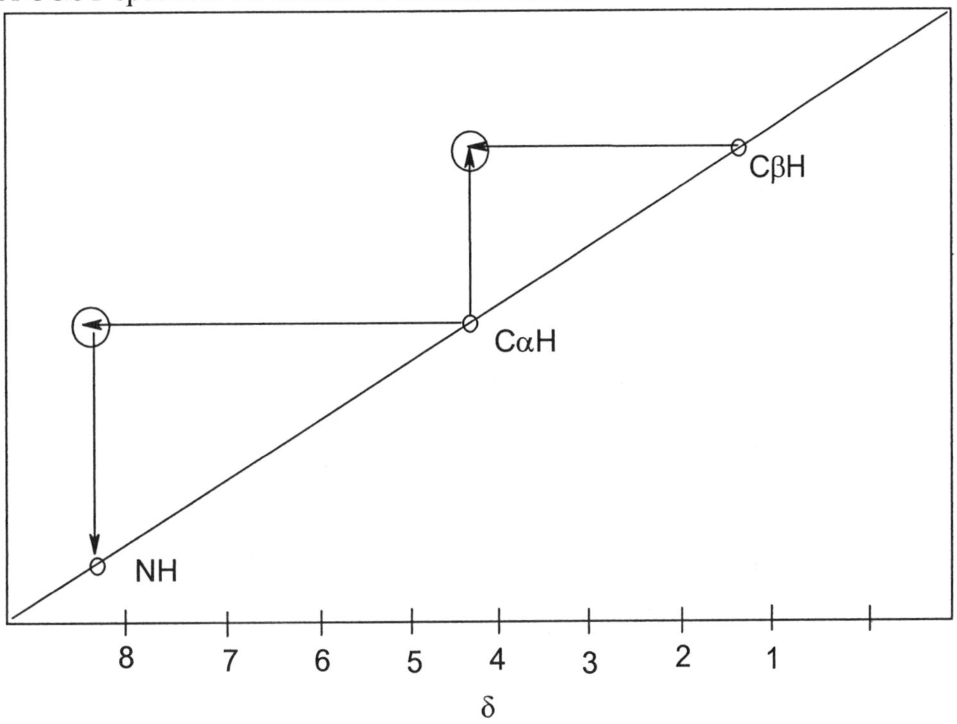

14.39 g for a free electron is 2.0023. In this molecular radical we determine the new g by $h\nu = g\,\mu_B B_o$. Solved for g:

$$(6.62608\times10^{-34}\text{ J s})\times(9.2231\times10^{9}\text{Hz}) = g\,(9.274\times10^{-24}\text{ J T}^{-1})(329.12\times10^{-3}\text{T})$$

$g = 2.002222$

14.40 The hyperfine coupling constant for each proton is 2.3 mT, since the difference between 330.2 mT and 332.5 mT is the same as the difference between 332.5 mT and 334.8 mT.

The g value is calculated below using the center line:

$$g = \frac{h\nu}{\mu_B B_0} = \frac{6.626 \times 10^{-34}\ \text{J s} \times 9.319 \times 10^{9}\ \text{s}^{-1}}{9.274 \times 10^{-24}\ \text{J T}^{-1} \times 332.5 \times 10^{-3}\ \text{T}}$$

$g = 2.002$

14.41 n nuclei of spin I interacting equally with unpaired electrons give:

The number of spectral lines = $2nI+1$

(a) for CH_3:

$n=3$

$I = ½$

4 spectral lines

(b) for CD_3:

$n=3$

$I = 1$

5 spectral lines

14.42 Using Equation 14.10, we can solve for the value of B_0 :

(a) At 9.302 GHz

$$B_0 = \frac{h\nu}{\mu_B g} = \frac{6.626 \times 10^{-34}\ \text{J s} \times 9.302 \times 10^{9}\ \text{s}^{-1}}{9.274 \times 10^{-24}\ \text{J T}^{-1} \times 2.0025}$$

$g = 0.3318\ \text{T}$

(b) At 33.67 GHz

$$B_0 = \frac{h\nu}{\mu_B g} = \frac{6.626 \times 10^{-34}\ \text{J s} \times 33.67 \times 10^{9}\ \text{s}^{-1}}{9.274 \times 10^{-24}\ \text{J T}^{-1} \times 2.0025}$$

$g = 1.201\ \text{T}$

14.43 The number of lines is given by $2nI+1$

Therefore, $5 = 2(2)I + 1$

$I=1$

14.44 (a) Figure 14.48 shows the EPR spectra of di-*tert*-butyl nitroxide radical at two temperatures. As the concentration of the radical increases, the electron exchange is predominant and the EPR spectrum is expected to broaden (similar to the one at 77 K). The spectrum at low concentration should look like the one at 292 K.

(b) The dependence of the EPR spectrum of nitroxides with temperature makes them ideal probes for lateral mobility in membranes.

14.45 The spectrum is three distinct pulses up-down to equal negative amplitude, then return to x axis. If temperature changes to low values (LN2) or high [M], movement will be restricted and distance between molecules reduced. Both effects will broaden the pulses.
High temperature (room temperature) and low concentrations (milli-molar) will increase the mobility and sharpen the spectral lines.